North-East India Tribal Studies

An Insiders' View

The Editors

Dr. Cheithou Charles Yuhlung completed his M.A (2002) and Ph.D (2011) from the Dept. of Sociology, North Eastern Hill University (NEHU), Shillong. His thesis is titled as '*Indigenous Religion of the Chothe of Manipur: A Sociological Study*'. He has worked as Research Associate (2011-12) in an UPE-Project, Dept. of Philosophy, NEHU, and also as Research Assistant (2006) in an evaluation team for an International NGO (IWGIA). He has published few papers in journals (Sociological Bulletin and IOSR) and in few edited books. He is the General Secretary of NEITRA and since 2012 he has been teaching in Women's College, Shillong.

Dr. P.G. Jangamlung Richard completed his M.A, M.Phil and Ph.D (2010) from the Dept. of Political Science, JNU, Delhi. His thesis is titled as '*Women in Post-Taliban Afghanistan: Socio-legal Perspective*'. He has served as Research Assistant (2009) in NCERT, Delhi. He also served as Project Officer and Research Associate (2012-14) in Dept. of Political Science, NEHU, Shillong. He is the President of NEITRA, and is currently teaching in St. Peter's College, Shillong.

North-East India Tribal Studies

An Insiders' View

— Editors —

Cheithou Charles Yuhlung

P.G. Jangamlung Richard

North East India Tribal Research Association (NEITRA)

2015

Regency Publications

A Division of

Astral International (P) Ltd

New Delhi 110 002

Cataloging in Publication Data--DK
Courtesy: D.K. Agencies (P) Ltd. <docinfo@dkagencies.com>

North-East India tribal studies : an insiders' view / editors, Cheithou Charles Yuhlung, P.G. Jangamlung Richard.
   p.   cm.
Includes bibliographical references and index.
ISBN 9789351306962 (International Edition)

   1. Ethnology--India, Northeastern. 2. India, Northeastern--Scheduled tribes--Social life and customs. 3. India, Northeastern--Scheduled tribes--Religion. 4. India, Northeastern--Scheduled tribes--Economic conditions. 5. Rural development--India, Northeastern. 6. India, Northeastern--Scheduled tribes--Politics and government. I. Yuhlung, Cheithou Charles. II. Richard, P. G. Jangamlung.

DDC 305.8009541   23

Published by   : **Regency Publications**
                          A Division of
                          Astral International Pvt. Ltd.
                          – ISO 9001:2008 Certified Company –
                          4760-61/23, Ansari Road, Darya Ganj
                          New Delhi-110 002
                          Ph. 011-43549197, 23278134
                          E-mail: info@astralint.com
                          Website: www.astralint.com

Laser Typesetting   : **Classic Computer Services**, Delhi - 110 035

Printed at   : **Replika Press Pvt. Ltd.**

PRINTED IN INDIA

Disclamation

All the papers herein reflect the author's view, and therefore, the *editors* have no claim nor any responsibility for any errors, mistakes and remarks mentioned. It is the sole author's core viewpoints from his/her experiences, observations and understanding that are presented herein. If anybody feels his/her sentiment is being hurt and unsatisfied with certain statement in the content of the paper one may consult and verify with the concern author to the corresponding address, email and mobile phone no. provided at the end of each paper.

Best wishes to all

Cheithou Charles Yuhlung
P.G. Jangamlung Richard

Dr. Vanlalnghak

Former Director, IGNTU, Imphal, Manipur, and
Professor of Philosophy & Culture
Department of Philosophy,
NEHU, Shillong - 22

Foreword

Tribals and Cultural studies has occupied very important and significant place in modern time. Almost all disciplines in academia and policy framers tries to involve themselves in one way or the other to contribute something. It is good to participate and tries to contribute something in burning issues like tribal studies wherein meaningful and scholarly contributions will render valuable authentic knowledge of the tribals and their cultures. But unfortunately, most often than not, many of these studies were carried out to suit short sighted vested purpose and in some cases studies were depended on secondary outsiders' view and has landed up in giving wrong picture of the tribal way of life. In this connection, it will not be out of place to mention that the participant's observation and insiders' view is important and authentic source of information.

This volume of essays – thematically broad in scope – has its focus on issues involved in understanding the complex concepts of culture from insiders' view is a good beginning. Some of the essays may required further theoretical and scholar articulation so as to enable to throw better light on issues dealt by them to the outsiders'. But I have no reservation on the authenticity of the text present in these essays as it is original to the extent in which it is the product of the actual participants. This will pave a long way in serving to the outsiders' in giving authentic knowledge of the tribals of the North-east India, their cultures, customs, religions, traditions, worldviews etc.

The *North-East India Tribal Research Association* (NEITRA) which brings out this valuable important volume is highly commendable in their endeavour. It is my pleasure to recommend to all readers' who are interested to understand tribals of the North-east India to read this volume pregnant with information. However, the final judgment is yours.

Dr. Vanlalnghak

Preface

The North-east India comprises of eight states; Arunachal Pradesh, Assam, Nagaland, Manipur, Meghalaya, Mizoram, Tripura and Sikkim. All these hilly states are predominantly inhabited by various tribal groups. According to the Census of India 2011, the total population of these eight North-eastern states is 4,55,87,982; while the total scheduled tribes is 1,26,79,736 and the scheduled caste is 30,30,415; whereas other cultural groups, especially from Assam, Tripura and Manipur comprises about 2,98,77,831. Each community big or small have its own distinctive history, language, culture, religion, world-views, marriage, economy and political systems, which is why the North-east region is also known as 'Cultural Hotspot' zone because of its 'cultural diversity'. Approximately 124 Scheduled Tribes are listed by the Census, besides hundreds of undefined sub-tribes in the region (some classified tribes overlaps with non-classified tribes since they are also found concentrated in other states). There are about 175 distinct languages (dialects) spoken and cultures that practices in this small hilly tropical zone.

Change, being a social phenomena and an inevitable force has affected most of the traditional social institutions and customs either in a small or big way. Thus, the region has undergone tremendous transitional changes from traditional to modern outlook beginning with the advent of Christianity and after merging with the Indian Union. In the areas of socio-economic poverty upliftment and political democratisation various discrimination and suppressive roles are also observed. As a result, various internal and external conflicts had occurred in many tribal societies with differing views both from outside and inside even on a single subject matter. This has confused and complicated the government in designing certain long terms policies and programmes for development.

Many young tribal academicians from North-east India, apart from thousands of common man seriously felt that they are being discriminated and subjected in almost every walks of their life; socially, academically, economically and politically because they are considered underdeveloped, backward, illiterate, barbaric, uncivilized and uncultured peoples from the hills. This judgment appears to be true from a different perspective looking from food habits, worldviews, religions, lifestyles, educations, etc. But on the other hand, most of the North-eastern tribes do have rich cultural heritage, sustainable economy, stable socio-political system, and high philosophy towards balance life who had live harmoniously and peacefully with nature for centuries. Now-a-days, the North-east people are more liberal and flexible in their outlook since the majority are Christians who have discarded their animistic religion, dogmatism and orthodox attitudes, and also by obtaining modern education. This allows them to accept and adapt to the new ideas and changed more easily than the mainland Indians or Arabs who are still subjected to the caste and Islamic systems.

It is also observed that some writings, despite efforts were put forth, were found with silly errors and unreasonable explanations since it was based on secondary and tertiary sources. The outsider seems to have written without even fully understanding or contacting the people as if they knew the people very well. This enraged many young tribal scholars for wrongly writing and misinterpreting about the people or failing to give the precise logical meanings of their culture and customs. For example, when some dynamic tribal academicians raise certain questions to clarify their doubts on such issues like their ethno-history, customary laws, causes of insurgency problems from certain writings that came across most of the Indian scholars could not provide correct answers, though some serious analysist manage to provide suitable answers and directions. Further, when challenged by these young tribal scholars for reasons of correction or rectification some made excuses for the reasons of being outsiders, time constrain, language/dialect, communication problems, ignorance, etc. which were not acceptable to the insiders.

Likewise, over the past decades, a multitude of policy prescriptions given by developmental experts for the region's socio-economic development shows unsatisfactory result due to lack of in-depth studies. All the drawbacks of developmental projects were pointed to the insurgency problems in North-east India without actually realising the causes, intricate culture and the sentiments of the people. It also shows a third party (who knew few truths) were involved in providing wrong information to the Govt. to formulate certain laws. Other problems include like delay in releasing funds, lack of transparency in the recruitment process, etc.

On the other hand, factual and sensitive political and economic issues that are constructive for society are often hidden from the public, on account of being a threat and problem to the ruling party's policies and programs, which actually were not. For example, the case of 'Tipiamuk Dam' where the inhabitants and down-stream people like Hmar, Zeliangrong and Bangladeshis were not consulted and evaluated about the damage and compensation during the survey. The third party (politicians) with little knowledge enraged the proposed project without proper evaluation, therefore has complicated when the project was launched.

Another current economic issue is the *Mahatma Gandhi National Rural Employment Guarantee Scheme* (MGNREGS), earlier known as *National Rural Employment Guarantee Act* (NREGA) meant for rural development is now echoed with corruption, mismanagement of funds, fake mutter rolls in many rural implemented areas resulting in chaos and mistrust among the villagers, despite bringing some positive changes and efforts to many areas. The other hot political topic is the 'Look East Policy 2020' in which many renowned scholars continue in the evaluative debates. The cross-fire statements share both positive and negative aspects.

The latest sensitive social issues (even seen in the news) are the rape cases, harassments and racial discrimination of the North-east people happening in Delhi and elsewhere in India. The North-east people are mis-identified and mistreated to be foreigners accusing to be Chinese, Japanese, Korean or Nepalese because of similar Mongolian features rather accepting as brethren Indians by birth and citizenship. When explained that 'we are also Indians from North-east' many mainland Indians don't know where 'North-east' is located. Negation of North-east ethno-history and maps in the school text book has resulting in poor geography for many mainland Indians. It is in such context that one may strongly argue that the North-east people were segregated, suppressed and misinterpreted about their identity, culture, customary laws, interest and grievances from simple to highly sensitive issues for many known and unknown reasons.

Hence, it is pertinent to question as to why even the experts failed to deliver the right decision. Another reason is that the writings of various scholars, military officials, journalists, official reports, traveller dairies, media persons and others who visited North-east wrote about the people and their problems mostly from external perspectives without obtaining much insights into the intricate customs and traditions of the indigenous tribal people of North-east. As mentioned earlier even some literatures on North-east are found to be non-participatory works, based on secondary and tertiary sources only (who gathered information from few informants to fulfil their assignments), except for few classical studies that need to be appraised and cherish. Since the data, views, sensitive reports did not present the reality of the situation the implementers failed to solve and bring various constructive changes to the region, despite drastic developmental programs are being made at different levels till today. The region is still undergoing the separatist and secessionist or unification movement among various major and minor ethnic groups, since each group and sub-groups (tribes and sub-tribes) differs from little to largely in terms of religion, languages, customs and traditions. Therefore, deeper understanding of the North-east people and their customs is a must if the social, education, economic development and political stability are to be brought in the region.

It is on such critical outset that some likeminded tribal scholars from the North-east states, irrespective of ethnic group, religion, discipline of subject and regional differences have voluntarily come together and formally launched the *North-east India Tribal Research Association* (NEITRA) on 9th December 2011, at North-eastern Hill University (NEHU), Shillong. Ultimately, it is felt that the people writes their own ethno-history and other situated problems to the existing literatures and enlightened us again their current ethno-history, custom and tradition, attitudes, religion, worldviews, language, marriage, economy and the political aspiration, environment, forest, wild-life conservation, including many other aspects. One of the basic objectives of the association enshrined in the motto: "Seek Truth to Progress" is to contribute in the field of academic. Based on the consensus, all the members agreed to come out with an edited book titled, *"North-east India Tribal Studies: An Insiders' View"*, for its first journal *v*olume. Members and paper contributors are exclusively tribals (for the moment) from the region who are pursuing Doctorial Research, Post-Doctorial Fellows, including Assistant Registrar, Assistant and Associate-Professors in various reputed universities of India and abroad, besides other professions.

The book contains fifteen chapters and is broadly divided into four sections: *Section – I*: Ethnography, Socio-Cultural and Linguistic Studies; *Section – II*: Religion and its Practices; *Section – III*: Rural Development and the Autonomous District Council for Hill Areas; *Section – IV*: Political Issues of North-east: Challenges and Strategies.

Therefore, all the papers herein reflect the author's viewpoint as an insider, speaking for the thousands of underprivileged people of his/her society for the betterment of the society. They represent the eyes, ears and voices of those people that cannot be seen or heard by other millions of peoples of the world. Since, the book is in the initial stage it invites serious papers from multi-disciplinary fields that are constructive and challenging for the society.

Best wishes to all

Cheithou Charles Yuhlung
P.G. Jangamlung Richard

Contents

Section – I
Ethnography, Socio-cultural and Linguistic Studies

Section – II
Religion and its Practices

Section – III
Rural Development and the Autonomous
District Council for Hill Areas

Section – IV
Political Issues of North-east: Challenges and Strategies

Section – I

Ethnography, Socio-cultural and Linguistic Studies

Chapter 1

The Bangru: The Lesser Known Tribe of Arunachal Pradesh

☆ *Tame Ramya*

ABSTRACT

The present chapter reports the ethnographic profile of an unknown or unrecognised small sub-tribe of Nyishi namely Bangru living in Sarli circle of Kurung Kumey district in Indian State of Arunachal Pradesh. A total of 15 villages from the circle were selected for the study. The data were collected using a set of ethnographic techniques *viz.*, observation, informal interviews with the villages and in-depth interviews with key informants the community. Within this study, I attempt to provide a general ethnographic outline of traditional Bangru society and culture as it existed a years ago when it was still relatively untouched by outside influences. My objective is to offer a systemic compilation of ethnographic data on traditional Bangru society, which may be helpful to those keen to know about Bangrus and to those interested in this region as any accounts of this land and its people are still not available.

Keywords: *Ethnography, Bangru, Sub-tribe, Nyishi, Arunachal Pradesh, Kurung Kumey.*

1.1 Introduction

The Bangru is one of the least-known sub-tribe of the larger Nyishi tribe of Arunachal Pradesh, with a population of about 2600 people. They inhabit mainly in the Sarli administrative circle of Kurung Kumey district (erstwhile Lower-Subansiri) in northern fringe of central Arunachal Pradesh, bordering the Tibet (China). They are spread in Sarli town and in few villages *viz.*; Bala, Lee, Lower Lichila, Upper Lichila, Machane, Milli, Molo, Nade, Namju, Palo, Rerung, Sape, Sate, Wabia, and Walu.

Till date, the Bangru is considered the sub-tribes of the larger Nyishi tribe of Arunachal Pradesh, although they differ in their origin and dialect. However, it is evident that both have somewhat similar socio-cultural specialities due to long association over the years. The major clans of Bangru are Pisa, Milli, Sape, Mallo, Tagang besides some minor clans. The Pisa clan is considered more advanced than its counterpart. The origin of the Bangru is not clear but it is certain that its origin is unparalleled with the Nyishi. They are similar in their physical appearance and are well versed in Nyishi dialect. They believed that they were the descendants of the children borne out of the *Ju* (Sun).

This study presents a conceptual framework on the historical development of the Bangru community of Kurung Kumey district. It is intended to provide a critical perspective. It describes how the Bangru maintained the elements of traditional culture in their day to day life and also attempts to textualise the oral history by incorporating their past perspectives.

This study centred on the Bangru who inhabits the Sarli circle of Kurung Kumey district. They are economically and educationally backward and also are deprived of many facilities enjoyed by other ethnic groups. It also focuses on the impact of modernisation on their culture and how their indigenous culture is in danger of extinction. This study has selected all Bangru inhabited villages of Sarli circle as the area of study.

1.2 Objectives of the Study

This study is ethnographic in nature and the basic objectives are:

  i. To know who are Bangru - their origin, migration from their oral tradition,
 ii. To provide a basic ethnographic understanding of their social institutions-religion, marriage, village administrative organisation,
iii. To understand their village economy and other livelihood strategies.

1.3 Methodology

The study is purely exploratory and descriptive in nature following the ethnographic model. Both primary and secondary sources of data have been used in the interpretation. Primary sources of data are obtained using various tools and techniques like household survey, in-depth interviews in the form of informal and unstructured questionnaire with key-informants of the community such as village elders, leaders, etc. and also by participant observation. Pelto and Pelto (1978) standard guidelines have been followed during the collection of the ethnographic data. Random sampling based on gender and age of the population had been followed. The universe of the study area confines to Bangru inhabited villages of Sarli Circle, in Kurung Kumey district of Arunachal Pradesh.

Since, there is any specific literature available on Bangru, the official records, documents and literatures available on the Nyishi in district headquarter of Kurung Kumey *i.e.* Koloriang has been used as secondary sources of data.

1.4 Universe of the Study

The Bangru a sub-tribe of Nyishi, spread over fifteen (15) villages in Sarli circle of Kurung Kumey District of Arunachal Pradesh constitute the universe of the study. Since, the Bangru constitutes unit of study basic information was collected through a comprehensive household survey schedule and using multiple data gathering devices, in order to make an in-depth investigation of the issues concerning people's participation and development of the target group. All the Bangru inhabited villages were selected because they all concentrated in these areas, although some populace have migrated outside the region of recently.

1.5 Bangru: An Ethnographic Profile

Bangru, an assumed sub-tribe of Nyishi inhabit the Sarli Circle, an administrative circle of former Lower Subansiri District, now part of Kurung Kumey District; live in what may be called as "*Ultima Thule* of Kurung Kumey District". Since, time immemorial they have met their subsistence requirements through mixed economic activities like agriculture (Jhum and Settled), hunting, fishing, gathering and other subsistence activities. Isolated in their remote, inhospitable, and high-mountain environment the Bangrus have had to find practical solutions to such basic problems as dearth of arable land for cultivation, lack of sufficient water for irrigation, and escalating population pressure on resources. The construction of terraces and traditional irrigation channels, a ritual complex that ensured the optimal use of seasonal conditions and limited time and space, and a system of communal and private land ownership that applied to this high-altitude region are some of the Bangrus' solution to their problems.

1.5.i Origin and Migration of the Bangru

Like most tribes of Arunachal Pradesh, the origin and migration of the Bangru is vague, since they have no written records. So far, no scholars or writers have ever made any earlier reference about Bangru even while documenting histories, customs and traditions of various tribes and sub-tribes of Arunachal Pradesh. This is probably due to the fact that no scholar have ever paid a visit to the land of Bangru inhabiting in one of the most remotest and inaccessible region of Arunachal Pradesh. What little is known about their origin and migration is based on their oral narration. However, it is believed that hundreds of years ago the Bangru migrated somewhere from Tibet and established their settlement around this place now called as *Sarli* and in its adjoining areas.

Mythological, the Bangru believes to have originated from a place called *Neto-Nello Puko*, meaning 'a place where the people fall/ came down from *Ludlu* (Sky/ Heaven) sent by *Aneya Ju* (Mother Sun), located in the present Sarli circle of Kurung Kumey district. Bangru, unlike many tribes of Arunachal Pradesh traced their descendant directly from the *Ju* (Sun). It may be noted here that there are two factions

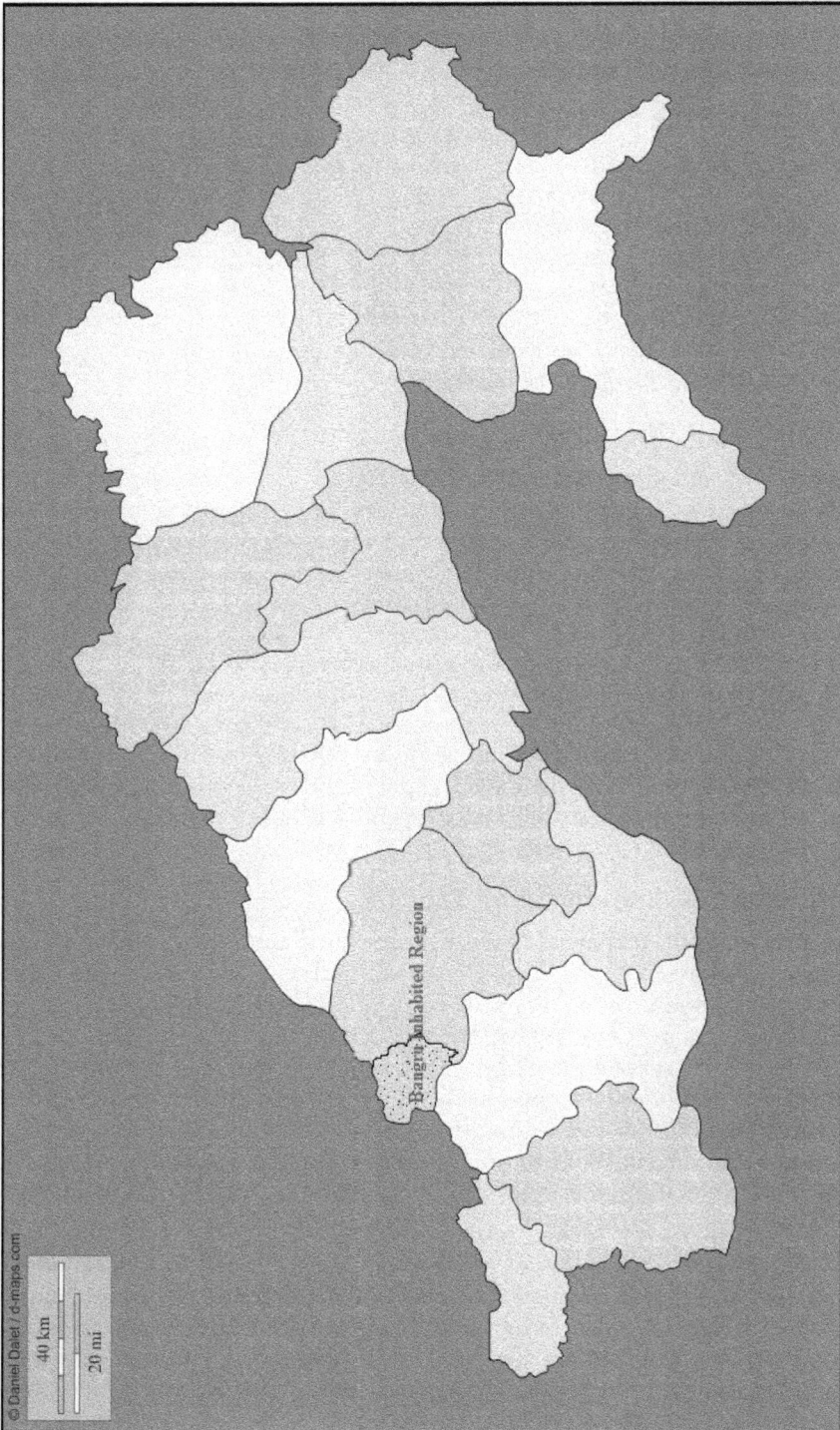

Figure 1.1: Map of Arunachal Pradesh Showing Bangru Inhabited Region.

among the Bangru divided on the basis of their versions of origin, *i.e.;* the general Bangru (*Phujoju and Milliju* and other minor groups), and the *Sape* who have a different version about their origin and migration.

According to the legend of the first group, over the years, Bangru had moved to areas nowadays called Sarli a place called *Neto-Nello Puko*. Some of the respondents claimed that earlier the word *Ju,* meaning 'Mother Sun' their ancestor name was used as suffixed in every of their clan's name. Hence, each clan were identified as *Phujoju* for Pisa, *Milliju* for Milli, *Malloju* for Mallo, *Tagangju* for Tagang, etc. But nowadays this term '*Ju*' have been commonly avoided by the people while identifying their clan's name.

Most Bangru prefers to call themselves as *'Taju-Bangru'*, but till date all the Bangru are known as a sub-tribe of Nyishi to the outsiders. So far, it cannot be ascertained as to when Bangru have entered to the present habitats. Some of the informants claimed that there is another branch of Bangru which they call them as *Wadu-Bangr* whom they believed has moved towards the western route *i.e.* the present East and West-Kameng districts.[1] So, they presumed that Aka and Miji (Sajolang) tribes are of common descendants under *Wadu-Bangru* branch. They substantiated their argument by comparing their *Phojoju* and *Milliju* with Miji's (Sajolang), *Rijiju* and *Khonjuju*.

On the other hand, the second group *i.e., Sape* clan traced their origin of migration from a place called *Jiila-Ralla)*. This group is considered the latter entrant to the present habitats who came to help the first group during warfare. These people assumed common descendants with *Memba* and *Khamba* who migrated towards east *i.e.* present *Mechuka* and *Tuting* regions of West and Upper-Siang respectively.

Thus, it is pertinent that the two groups are of different origin and from different place despite now speaking similar language, customs and tradition, since both have different versions of origin, migratory routes and different ancestral history. This assumed that in the past, when they migrated to the *Sarli* region either the earlier group or the latter group must have suppressed and dominated the other, thereby the weaker must have submit and assimilated with the dominant group under certain socio-economic and political conditions.

1.5.ii Population Composition

Bangru, numbering about 1,023 (39.35 per cent), out of approximately 2,600 persons in the Sarli circle is one of the least-known tribes of Arunachal Pradesh, located in Sarli Town of Kurung Kumey district.[2] There are not more than 15 villages. Although there is no separate Census record on this community, however, according to the data gathered by the author, the Bangru accounts for about 1.14 per cent of the total population of Kurung Kumey district. The Bangru population of the Sarli Circle is given alphabetically according to their village.

1.5.iii Language

Bangru is an unclassified linguistic group which was earlier included in the Upper Assam language group of Tibeto-Burman language family, though no evidence

Table 1.1: Distribution of Bangru Population in Different Villages

Sl.No.	Village	Total Population	Male	Female
1.	Bala	10	5	5
2.	Lee	64	36	28
3.	Lichila (Lower)	72	32	40
4.	Lichila (Upper)	54	29	25
5.	Machane	65	35	30
6.	Milli	102	45	57
7.	Molo	22	12	10
8.	Nade	12	5	7
9.	Namju	33	18	15
10.	Palo	30	11	19
11.	Sape	152	75	77
12.	Sarli Town	306	134	172
13.	Sate	28	7	21
14.	Wabia	66	24	42
15.	Walu	6	2	4
	Total	**1023**	**471**	**552**

Source: The Electoral Registration Office, Koloriang.

is available on it language affiliation. It is different from languages of Nyishi and Puroik. But the three groups have socially and culturally very close affinity among each other. However, it is worth noting that the Bangru language has been largely influenced by Nyishi and as a result changes have occurred in the internal reconstruction of Bangru speech forms. But it would not be wrong to opine that the language of Bangru is remarkably pure. Due to intermingle of Nyishi, Bangru and Puroik languages there reflects some affinities in their verbal communication. There is no evidence as to put the Bangru language into some linguistic diverse group and subgroups since no specific study on linguistic affiliation of the Bangru language is done, so far. The language shift is observed among the people where they inclined more to Nyishi, a neighbouring tribe of the region.

1.5.iv The Settlement Pattern and Housing

The Bangru settlement is small in terms of its size and population. As per my field records, the total population of the Bangru villages varied between 6-306 persons, the highest being the population of 306 persons observed at Sarli Town, while the lowest is 6 persons in Walu village. This is because a person is living with his Nyishi relative in the village. It is also to mention here that each settlement is recognised as village by the state government but for administrative convenience 2-3 settlements are pooled together to make a complete single village where one representative in the form Anchal Samti Member (ASM) is elected. However, a typical Bangru settlement usually consists of 50 to 100 populations. Each of these small settlements inhabited by about 10-20 households is known as *Neye* (village). It is also found that 5-10 hamlets on a particular hill are combined and given a village name.

The typical Bangru settlement is characterized by sparse distribution with little disorderliness. The houses are distributed unevenly in the settlement area. Many a time, the houses are constructed wherever a place is available. Traditional households in the Bangru settlement are thatched and uniform in their structure. However, with the passage of time, such traditional thatched houses are now being replaced with modern day's CGI sheets (tin sheets) to whom people deem more comfortable and secure to use. When the sons separate from their parents usually after marriage, they construct another house adjacent to their parent's house. If there is no space available, they may construct at some other place nearby.

1.6 Social Organisation of the Bangrus

Bangru society is patriarchal with a distinctive character of tribal endogamy and clan exogamy social system. We have already noted that the Bangru tribe is dividing into five clans and each clan few minor clans. The Bangrus' tradition is unanimous in talking of five clans although the number of sub-clan differs. The Bangrus accept the rule of the clan system and the myths, which form its background, are a key to the understanding of almost everything that is distinctive in their way of life. Violation of tribal endogamy and clan exogamy are the crimes in the Bangru society and those who break these rules are deal with exemplary penalties. The fundamental and primary feature of social organisation is represented in every Bangru village. The presence of different clans in a village demonstrates obviously the democratic character of Bangru society.

The Bangrus constitute a well-defined and homogeneous group of people. Although their villages are scattered over a wide area, the Bangru people everywhere speak the same language and follow the same customs, have the same traditions, beliefs, rites, and ceremonies. Such small differences as they present from place to place are hardly greater than those obtaining between the villagers of adjoining regions. All are bound together by a common sentiment for the tribal name, reputation, tradition, and customs. At least five clans of Bangrus, each bearing a distinctive name, are recognised. The word *Neye*, which appears in the names of each group, means village or settlement, and it seems probable that these five clans represent 15 original Bangru villages *viz*. Bala, Lee, Lichila (Lower), Lichila (Upper), Machane, Milli, Molo, Nade, Namju, Palo, Rerung, Sape, Sarli Town, Sate, Wabia and Walu which have contained the whole Bangru population.

1.6.i *Gyaiidya* (Marriage) Among the Bangru

Marriage in Bangru society, as in all other societies is a turning point in the life history of an individual from where he branches off from the parental roof and establishes a new unit. A girl on her marriage abandons her parent's home and goes to live with her husband. Traditionally, polygamy was prevalent in Bangru society where number of wives was the symbol of being wealthy in the society. However, in contemporary Bangru society monogamy is preferred because family problems arise due to multiple wives. The cross-cousin marriage system that is marriage with mother brother's daughter is preferred form.

It is worth mentioning here that the Bangru, Nyishi and Puroik inter-marry frequently and have developed close socio-cultural relations. These inter-marriages and close socio-cultural contacts have reduced much of their differences, although some dissimilarity prevailed. Marriage among the Bangrus generally involves the following considerations:

  i. Enhancement of social and economic status in the society.

 ii. Addition of working hand in the *Wua* (fields).

iii. Housekeeping partner.

 iv. Meeting the biological and psychological needs.

  v. Procreation.

 vi. Financial gain to the bride's in-laws.

vii. Increase in sphere of influence and cooperation through new relationships.

Among the above marriage considerations, priority is given for procreation, to meet the biological and psychological needs, for helping in domestic works and housekeeping.

Marriage in Bangru society is traditionally arranged by the parents preferably with the people of equal social status. The Bangru do not prefer marriages between specific kinsmen, for in olden days the alliance of the two families called for mutual support in fends and raids (Haimendorf 1982:64).[3] Matrimonial alliances were a means to gather allies to defend itself against attack from enemies. Marriages are also arranged to obtain some valuables of fine quality and repute which the girl's family may be having.

As in many tribal societies of Arunachal Pradesh, the concept of divorce in Bangru society did not have much significance. Usually, divorce is not sought by a man. When he does, it is not binding on him to return the gifts given to him in exchange for the bride price. In lieu, a fine may be imposed for deserting the wife. But, if it is initiated by a woman, her parents are obligatory to return the bride price, sometimes double of the actual amount. Customary law does not speak of any divorce alimony or compensation for a divorcee.

1.6.ii Family (*Lameii*)

The family is a universal institution and has existed throughout the history of human society. It is the smallest social unit consisting of parents and their unmarried children. It is the simplest and the most important primary group in society.

Lameii (Family) in the Bangru society is the outcome of marriage. Bangru has the tradition of joint family system but in present society nuclear type of family which is most commonly found. This is because married sons more often tends to live in different house by their own resulting in the nuclearisation of the family. With the passage of time, family has undergone changes gaining and losing valour shapes and characteristics. The present age of economic development and cultural revival have posed some new challenges to the institution of family; leading to radical changes in the structures and functions of family. Unlike in the past, father does not have control over the whole family in many aspects. Sons who live separate from their parents make their own decisions on behalf of their families. However, the institution of family is surviving and will survive for the survival of the society itself.

With regard to family inheritance, only males are able to inherit real property on a permanent basis, although among the Bangrus, for example, a widow might be granted a temporary inheritance of her husband's property (on her death it would pass to her sons), and a daughter(s) receives ornaments in the form of *Tate* (beads) and other such traditional ornaments. Similarly, property is often divided during a man's lifetime, with each son receiving a portion of the property from his father upon his marriage. The property of any son dying without male children during his father's lifetime reverted to the latter, and after the father's death it went to the youngest son.

1.6.iii The Clan

The basic feature of social organisation depends on the division of the community. In Bangru there are two distinct endogamous groups; they are *General Bangru* and *Sape.* Within *General Bangru* exogamous is still prevailing, which means a man from a particular clan marries a woman from a different clan, although both of them belong to same group. For example, a member of *Pisa* clan can marry a girl from *Milli* clan though both the clans are from within *General Bangru* group. Each clan identifies a village or group of villages and takes the names of that place they presently inhabit. The *General Bangru* group is divided into four major clans like *Pisa, Milli, Mallo (Mullo/ Mullong), Tagang,* and few minor clans, while *Sape* group is considered to have only one clan with the same name of the group.

1.6.iv The Village

Anthropologists, from the very beginning have studied the village as an autonomous human institution from the view points of political institutions, social interactions, inter-personal and inter-family relationships, etc., and have come to the conclusion that the village, despite various influences and changes, has retained its particular traditional institution. The Bangru village, being the most traditional and ancient institution, crystallised a whole system of social, political, and ritual structures. The clan is the most important social unit in Bangru society. The traditional polity is based upon the village. The village is a territorial unit claiming an exclusive right to a tract of land with clear boundaries.

On certain occasions the village drew a strong spirit of cohesion from its members. There is considerable local patriotism based upon a host of legends and colourful history of the village's past exploits. Another important aspect of the village is its function as a unit of defence. Most villages (or its member clans) are sometime in state of feud, so there is a need for perpetual alertness and vigilance, for a strong defences that would enable the villagers to resist attack without inordinate difficulty or great loss of lives. For this reason many villages are built on the hill tops, where a perennial source of water is available nearby the site. The villages are permanent and are encased in extensive and impressive circumvallation, now represented by the efforts and innovations of many generations.

1.6.v Kinship (Bangru *Guii-Koro*)

Kinship system is usually seen as a method of organising marriage relations between groups. Through marriage, members are recruited to kinship groups. The kinship is helpful to study the means of genealogies. The Bangrus are having a well planned kinship and which has got some interesting features also. The most important feature is the use of the same term *Alo* for grandfather on one hand but on the other for father-in-law (of both man and woman). Another feature may be found in the using of term *Ako* and *Mesebya* for all brothers and sisters (elder or younger) respectively. *Mesebya* is also used to refer to father's sister. The important features of the Bangru kinship are the system of existence of two well-marked groups of terms expressing bonds of kinship. Similarly, with regard to grandfather (*Alo*) and grandmother (*Asse*) respectively they used to refer to both paternal and maternal grandfather and grandmother. Further, the word '*Asse*' is also used to mean either grandmother or mother-in-law or wife's mother too, which signify terminologies are limited.

Another interesting pointing is that the Bangru system has two set of kinship terms, those used in direct address and those used when speaking of relatives who do not correspond closely with one another. This system distinguishes widely between elder and younger member of the family and clan. The Bangrus never mention their *kiini* (mother's brother) and *alo* (wife's father). Similarly, a man is also prohibited in mentioning the name of the father-in-law and mother- in- law (*asse*). The Bangru terms of address and their equivalent words in Nyishi and English are given in the Table 1.2. I have included the Nyishi here only to find out the similarities and dissimilarities of wordings within them.

1.7 Political Organisation of the Bangru

In parallel with the other tribes of Arunachal Pradesh, Bangru also have certain legal system *i.e.* their customary law. Bangru customary law in its traditional manifestations is not embodied in any formal codes. There are no written laws, although many rules of conduct are epitomised in proverbs and kindred sayings. There is no well-defined corpus of legal maxims and principles. The laws of the Bangru are to a considerable extent inherent in their social systems. They exist as rights and duties developed through the course of time out of man's efforts to adjust his behaviour in relation to his fellows and to the physical environment he shares and exploits them. They have accepted from the very nature that if they can satisfy the fundamental and common needs of the society, the more it is binding and obligatory.

The traditional political system of the Bangru cannot be seen in segregation from its culture, tradition, economic and social life. The Bangru village is a unit of village-state having independent governance in many activities. In actual term, Bangru does not have specific political form rather it shows parallelism with Nyishi's *Nyele*, a form of council that performed the judicio-administrative functions of a village or community.[4]

Table 1.2: Bangru Terms of Address and their Equivalent Words in Nyishi and English

Bangru Terms	English Equivalent Words	Nyishi Terms
Achowa	Mother's Sisters (common)	Amu/Mu
Aneya	Mother	Ane
Ako	Brother (common)	Achi/Abang/Buru
Alo	Father-in-law	Atu
Asse	Mother-in-law	Ayu
Juchobii	Sister's Son and Father's Sister's Son	Ku
Juchobya	Sister's Daughter and Father's Sister's Daughter	Ku
Kiini	Mother's Brother	Akh/Kiigh
Mechemya-Nyiib	Grandson	Ku-Nyaga Huiish
Mechemya-Nyiiwai	Granddaughter	Ku-Nyeme Huiish
Melgya	Husband	Nyulu
Mesebya	Sister (common)	Anyi/Barme
Mii	Wife	Nyahang
Miibi	Father	Abu
Miibo/ Miwo/Mibow	Daughter's Husband and Son-in-law	Magbu/Magtey
Minyii	Son's Wife and Daughter-in-law	Nyaahang
Muju-Nyiib	Son	Ku-Nyaga
Muju-Nyiwai	Daughter	Ku-Nyeme

Source. Fieldwork (July 2011)/Bangru Villages.

The village council is usually comprises of Gaon Burahs (GBs) from some selected village elders who are expert in dealing with certain types of cases or disputes.[5] The role of the council members is to settle all disputes amicably and impartially between both the parties. The council members enjoy a special status in the society. But they do not receive any remuneration for their status or rank except getting a share of the fine imposed to the accused. The most common punishment is the imposition of a fine, generally live stocks. The amount varies, according to the nature of offence and position of the offender, his previous record and his ability to pay by a single *su* (mithun) or his property that can be confiscated.

In the process, the role of *Gingdung* (mediator) is highly commendable. The *Gingdung* is a person who plays a very vital role in mediating between two parties to come into a consensus on any case or dispute. It is only because of his ability that the two parties accept for discussing the issue in front of the council and ordinary villagers.

In the house of council, the wrong doer or accused person is brought before the members of council to hear its decision. Almost all cases or disputes are settled in the council except some heinous crime like murder. Generally, the village council settled cases like marriage dispute, land dispute, adultery, theft, widow-remarriage, etc. by abiding their traditional customary law.

In contrast to the earliest norms of social justice dispensation in the Bangru society, the roles of village council and Gaon Burahs (GBs) are diminishing. The factors behind such changes are like the introduction of modern electoral political system in the form of Panchayati Raj system and the coming of Christianity. Nowadays, people approach directly the village panchayat as they have rights framed in rules and regulations for the welfare of the villagers. On the other hand, those Christian converts have neglected their customary law and did not approach the village council for settling any dispute.

However, there is no instance of influencing the functioning of the village council by any political parties. They are not able to influence the functioning, as the political parties are mostly seasonal. Government is taking steps to stimulate the village council by announcing allowances and honorarium for Gaon Burahs (GBs). This seems to be a kind of reorganisation of the village council. On India's Independence Day (2012), the district administrator recognises the Gaon Burahs (GBs) by awarding red-coat and other honorarium.

1.8 Livelihood Strategies of Bangru

The socio-economic status of the inhabitants of these areas varies greatly due to a number of factors. For examples transportation and communication facilities, agricultural practice, access to various modern resources related to development, culture contact with other neighbouring communities, etc. In the past, the Bangru people live in the group, mostly residing in close proximity to forest and river fringe because their life is depended on the forests and rivers (as told by Mr. Milli Takang, a key informant). This was due to their dependency on forest products like firewood, access to wild medicinal plants for curing diseases and for fishery being their main source of livelihood. They have started practicing shifting cultivation. They practiced hunting wild animals and gathering wild fruits and herbs. The Bangrus are expert in capturing wild animals, perfected in hunting and fishing. They used various traps and tools and natural poisons while fishing in rivers and streams. They have generated enormous knowledge on a large number of plants species on which they have depended for centuries. Due to this, forests were the most important resources for them in terms of food, fibre, medicine, housing materials, fodder and various other needs.

Agriculture is the most important economy of the Bangru. They depend upon their plots not only for the major part of their sustenance, but also for a cash income since few years back. The staple crop cultivated by the Bangru is *eay/ eaii* (paddy). Other crops like maize, *tamai* (millet), finger-millet supplements rice at regular interval. There is a tree called *Lavo* known as 'famine food' which supplement their foods at very difficult times like during famine often occurred due to bamboo flowering.[6] This tree is particularly predominant among the neighbouring Puroiks, for whom, it is serving as staple food.

The Bangru shifting cultivation is carried out in this way; a plot of land is planted with paddy, maize and other eatables for one year. In some of the villages, plot is allowed to lie fallow for 4-5 years, after which it is again put back into cultivation. In theory, the cycle of fallow and carrying crops continues indefinitely. But the Bangru people conceived that some plots "tire" more quickly than others, therefore must be allowed to rest after few years of used.

1.9 Religious Beliefs and Ritual Practices

Belief system and ritual practices play an important role in the religious life of the Bangru. They strongly believed in certain supernatural beings able to influence the destinies of the living either for good or for evil. Their basic religious life is quite similar with those of their neighbouring tribes like Nyishi and Puroik. Like other tribes, they have also developed myths of creation; the Sun and the Moon; the origin of man and about death. The concept of soul which they called *arey*, is that moment it occurs when the spirit is separated from the body at the time ones death.

According to Bangru, their religion is polytheistic in nature meaning they believed and worship multiple deities, usually assembled into a pantheon of gods and goddesses. Their traditional religion is *Donyi-Polo* or *Donyi-Poloism*.[7] They believe that some trees, stones and hills are the abodes of the spirits. Like all animistic religions, the Bangru belief consists of multitudes of benevolent and malevolent spirits. To some it is attributed to the creation of the world, to others the control of natural phenomena where the destinies of man from birth to death are governed by a host of divinities whose anger must be appeased by sacrifices through ritual ceremonies.

1.9.i The Bangru Religious Beliefs

The whole belief systems of Bangru religion may be confined to as:

1. Belief in Supernatural Powers.
2. Belief in the Supreme God.
3. Belief in Spirits World.
4. Belief in Magic and Witchcraft.
5. Belief in Worship, Prayer and Sacrifice.
6. Belief in Life after Death.
7. Belief in Soul.
8. Belief in *Tyameii* (Dream).

1.9.ii The Bangru Religious Practices

The Bangru religious practices are inherent in their daily ritual performances such as:

i. Ritual to Vindicate the Truth.
ii. Ritual related to Agriculture.
iii. Ritual related to Evil Practices.
iv. Ritual on Human Death.
v. Ritual for Immediate Healing.
vi. Ritual related to Creation.
vii. Ritual related to Purification and Divine Favour.

1.10 Bangru Festivals

Like many tribal societies, the Bangru also have their festivals. The complexity of many clans and their interlocking nature are crucial aspects of their festivals, and is notable when found at the Bangrus' technological level of development. The major agricultural festival of the Bangrus is similar to Nyishis of Koloriang area where they celebrated *Longte-Yullo* or simply *Longte* or *Lungte* as major festival. More interesting fact is that the myth behind this festival is entirely of Nyishis; and Bangrus have no independent myth of the origin of this festival though they are entirely a different ancestral group.

Literally, *Longte* means a large wooden barricade/ fence, which is erected on community basis with a belief that it demarcates the domain of human beings and spirits from ill-intended trespasses. This festival is celebrated on the advent of spring season in the month of April (*Lachar-Polu*). The festivals, an aspect of Bangru religion in the broadest sense, are ceremonies that instil, especially in the young, profound feelings for and beliefs in the Bangru way of life. The festivals provide guidelines for acting out traditional roles, thereby sanctioning them, as well as social settings in which the individual can experience joy and express love. The festivals contribute in this way to the high social cohesion that is characteristic of the Bangru socio-cultural system.

1.11 Christianity among the Bangrus

The study of the Bangrus' traditions and institutions would be incomplete without taking into account the influence of Christian missionaries. Christianity has a significant influence upon the Bangrus and apart from missionary activities they undertook many social welfare activities for the people. Actually, Bangrus conversion to Christianity started in the latter part of last the century when Baptists Christian Missionaries under the aegis of *Kurung Kumey Baptist Mission Field* (KKBMF) set their foot in Bangru inhabited regions whereby other missionaries began to work. The first people to encounter Christianity were those settled in Sarli town. Unlike in other part of Arunachal Pradesh, the Bangru people were not forbid to profess this eccentric religion. There was no any sort of prosecution against those who convert rather the people acclimatized themselves with their own will. This may be one reason why Christianity already flourished, especially among the neighbouring Nyishi tribe.

With the influence of Christianity, they have shifted from their old beliefs of *Ju-Libaying* (*Donyi-Poloism*) to Christianity but they mostly attend their basic traditions and social customs. The recent exposure to western and Christian ideas, especially education and religious practices, has resulted to a change in the Bangru way of life. This change caused a division into traditionalists and non-traditionalists. The traditionalists regarded themselves as the custodian of tribal customs while the non-Traditionalist view bringing about changes to their society by adopting Christianity or Westernisation. It is however, important to note that these groups are not geographically distinct, but lived side by side within their community and shared many social aspect of life. It is important to note that despite being Christians some are still interested in maintaining their typical Bangru traditional way of life.

1.12 Conclusion

Most tribal societies have its own distinctive characteristics, socio-cultural value-system, traditional norms and mores, religion, political, economic and kinship systems. It has its own approach to life and death, disease and sickness, individual and community, and above all a sense of identity. This sense of cultural identity or image always has positive and negative facets. It defines the traits of solidarity and uniqueness of the tribal group, and shows the differences from other groups from the perspective of the larger society.

The study on *'The Bangru'* is immensely important from the sociological and anthropological points since no study has ever undertaken before into their social systems and other cultural aspects. This ethnographic piece of writing on Bangru is a humble attempt to highlight their earlier history and culture of the people which is not available as yet. Today much has changed from the past and is still changing, confronted with new social, political, and economic circumstances. Since, this is the first study on Bangru, there are difficulties in drawing conclusions despite the empirical data collected through various ethnographic methods and techniques. But, I have taken great care in recording the data of my participant and non-participant observations to be as accurate as possible.

1.13 Suggestions

Some suggestions may be made on the Bangru tribe of Arunachal Pradesh based on the field study: They are:

1. Expansion of Employment Opportunities

Some of the traditional occupations of Bangru are basket making, fishing, hunting, felling trees, etc. Now-a-days, these works have not much demand in the area as industries and modern equipments have replaced them. At this juncture the Government needs to take some necessary steps to provide employment opportunities for them. Now, some items which they produce have no market, besides there are also middle man's exploitations. In this regard, the Government and Non-Governmental Organisations (NGOs) need to make certain schemes and plans for proper marketing to enable them to export their arts and handicraft products, thereby benefiting them economically and elevate their poverty.

2. Need Improvement for Better Transport and Communication Facility

Road transport and communication facility are the need of an hour if development is to be brought to Bangru's area, since most of their villages are located in the interior remote places. Lack of proper transport facilities has limited their knowledge about their rights, benefits, and chances for good job, education and medical care. They have to walk a long distance to the nearby small towns like Sarli and Koloriang for anything and everything like medical, educational and other governmental assistance purposes.

3. Need Improvement for Basic Amenities

Since, Bangru are located in far-flung remote villages they lack most of the basic amenities. Global issues or political issues are not their main concerned rather everyday survival such as food, housing, health and clothing are their main worries. Though the Government provide certain schemes for their basic needs unfortunately these are not becoming reality anymore. Thus, the Government and NGOs should set up an agent in accessing and monitoring such schemes to avail to them.

4. Need Emphasis on Education

Since, most Bangru are living in remotes area and suffers from various kinds of depravities (social, economic and political), education is the only solution that will help them to overcome all of these depravations. Most of the schools in the villages are in myriad of problems like lack of infrastructures, textbooks, etc, qualified teachers. Many Bangru children are school dropouts, since their schools are located far and also their poor economic backwardness.

5. Need Improvement for Health Facilities

Since, most Bangru live below the poverty line they lack proper hygienic food, and thus suffer from various kinds of diseases and anaemic sicknesses. Though few organisations are working among them there is no satisfactory development with regards to their health related problems. Many villagers are addicted to the bad habits of alcoholism, smoking and chewing tobacco too. There is only one primary health centre located at Nade (New Sarli), that too, without sufficient medical facilities like medical staffs, medicines, etc. There is a great need to spread health awareness programmes by trained social workers and also to establish more health centres in the region.

6. Need Encouragement for Ethno-medicines, Arts and Handicraft Items

Traditionally, Bangru are rich in ethno-medicines, arts and handicraft knowledge but with the advent of Christianity and modernisation such traditional knowledge are vanishing each day. Some of their local products need a proper marketing. The Bangru are rich in their ethno-medicines plants which can be, identified, trapped and even commercialised.

7. Need for Empowerment of Women

In contemporary Bangru society women are treated as subordinate to men. The Bangru women are the most sufferers in their community, since most of the domestic works and other responsibilities are shouldered by them. They took care of their children, went to the field (*wua*) and gathers food items. In order to uplift the women status in society they may be empowered by providing education, job oriented training programs, employment, encouraged to form women's society and also emphases on their rights related to equality, property, protection and other beneficiaries.

Thus, the researcher felt that if the Government and some NGO's can make some committal intervention in the focus areas of socio-cultural, education, economic and political systems in lives of this little unknown Bangru tribe of Arunachal Pradesh, perhaps there could be much development in their society. Thereby, it is expected that there can be improvement in their living standards, income, literacy rate, health and lessen their socio-economic and political problems.

Acknowledgements

I would like to thank all the informants for their time and commitment, without their participation this paper would have not been possible. Further, I would like to thank my guide and supervisor Prof. S.K. Chaudhuri, for his guidance and constant support including field visits for actualising this article.

Endnotes

1. There is no evidence on the relations of the first group with Aka and Miji and second group with Memba and Khamba. The information is entirely in accordance with the narrations of the informants.

2. This is in accordance with the recent revised electoral rolls carried out in the month of January, 2011. This is exclusively the population of persons having 18 years of age and above. No separate population statistic for Bangrus is available yet.

3. This was refers to the Nyishi society to whom Bangru society have basic similarities in their marriage customs.

4. Nyele, in actual sense of the term is the whole process of diplomacy in justice dispensation through a Gingdung (mediators).

5. Gaon Burahs here refers to those who have red-coat with modern judicio-magisterial powers.

6. Lavo is was the staple food of the Bangrus before they shift for contemporary modes of cultivation. The term is called as Tasse and Rangbang among Nyishis and Puroiks respectively.

7. These terms in Bangru language are known as *Ju* for Sun and *Libaying* for Moon respectively. Since Bangru does not have a religion with their terms; so the *Donyi* and *Polo* is being used.

References

Berreman, G.D. 2004. 'Ethnography: Method and Product', in V.K. Srivastava (Ed.) *Methodology and Fieldwork*. New Delhi: Oxford University Press.

Bernard, H. 2002. *Research Methods in Anthropology: Qualitative and Quantitative Approaches*, (Third Edition). New York: Altimira Press.

Bora, D.K. 2000. 'Traditional Nishing Religion and the Change', in M.C. Behera (Ed.) *Tribal Religion: Change and Continuity*. New Delhi: Commonwealth Publishers.

Chaudhuri, S.K. 2003. 'The Anthropology of Lesser-Known Tribes', In T.B. Subba and G.C.

Ghosh (ed.). *The Anthropology of North-East India*. New Delhi: Orient Longman.

Choudhury, S.D. 2008. *Gazetteer of India: Arunachal Pradesh* (Subansiri District). Itanagar, Govt. of Arunachal Pradesh: Himalayan Publishers.

Das, S.T. 1986. *Tribal Life of North-eastern India: Habitat, Economy, Customs and Traditions*. New Delhi: Gyan Publishing House.

Dutta, P. and S.I. Ahmad,(Eds). 1995. *People of India: Arunachal Pradesh.* Kolkata: Anthropological Survey of India, Seagull Books.

Elwin, V. 2006. *A Philosophy for NEFA*. Itanagar: Directorate of Research: Arunachal Pradesh.

Fetterman, D.M. 1998. *Ethnography Step-by-Step* (Second Edition). Thousand Oaks, CA: Sage Publications.

Fortes, M. and E.E. Evans-Pritchard. 1940. *African Political Systems*. London: Oxford University Press.

Furer-Haimendorf, C.V. 1950. *Ethnographic Notes on the Tribes of the Subansiri Region*. Shillong: The Assam Government Press.

Furer-Haimendorf, C.V. 1962. *The Apa Tanis and their Neighbours: A Primitive Civilization of the Eastern Himalayas*. London: Routledge & Kegan Paul.

Furer-Haimendorf, C.V. 1982a. *The Highlanders of Arunachal Pradesh*. New Delhi: Vikas Publishing House.

Furer-Haimendorf, C.V. 1982b. *Tribes of India: The Struggle for Survival.* California: University of California Press.

Garson, J.G. & C.H. Read, ed. 1892. *Notes and Queries on Anthropology* (2nd Edition). London: Council of the Anthropological Institute.

Hammersley, M. and P. Atkinson 1995. *Ethnography: Principles in Practice* (Second Edition). London: Routledge.

Hasnain, Nadeem 1991. *Tribal India*. Delhi: Palaka Prakashan.

Jairth, M.S. 1991. *Tribal Economy and Society*. New Delhi: Mittal Publications.

Jha, S.D. 1988. *Socio-Economic and Demographic Dimensions of Arunachal Pradesh*. New Delhi: Omsons Publications.

Kar, K.R. 2003. 'Tribal Social Organisation', in T.B. Subba and G.C. Ghosh (ed.), *The Anthropology of North-East India: A Textbook*. New Delhi: Orient Longman Pvt. Ltd.

Mair, Lucy. 1965. *An Introduction to Anthropology*. Oxford: Clerendon Press.

Maitra, A. 1993. *Profile of a Little-Known Tribe: An Ethnographic Study of Lisus of Arunachal Pradesh*. New Delhi: Mittal Publications.

Majumdar, D.N. 1950. 'The Affairs of a Tribe: A Study in Tribal Dynamics', in *The Ethnographic and Folk Culture Society*. Lucknow: Universal Publishers Ltd.

Miri, S. (ed.) 1980. *Religion and Society of North-East India*. Delhi: Vikas Publishing House Pvt. Ltd.

Mishra, N. 2004. *Tribal Culture in India*. Delhi: Kalpaz Publications.

Mohanty, P.K. 2004. *Encyclopaedia of Primitive Tribes in India* (Vol. 1 & 2). Delhi: Kalpaz Publications.

Nath, J. 2000. *Cultural Heritage of Tribal Societies* (Vol.1 & 2). New Delhi: Omsons Publications.

Olson, J. S. 1998. *An Ethnohistorical Dictionary of China.* Westport, Connecticut (USA): Greenwood Press.

Pandey, B.B., D.K. Duarah and N. Sarkar, ed. 1999. *Tribal Village Councils of Arunachal Pradesh.* Itanagar: Directorate of Research.

Pelto, P.J. and G.H. Pelto. 1978. *Anthropological Research: The Structure of Inquiry.* London: Cambridge University Press.

Raatan, T. 2004. *Encyclopaedia of North-East India:* Arunachal Pradesh, Manipur and Mizoram, (Vol.2). Delhi: Kalpaz Publications.

Rikam, N.T. 2004. 'The Faith and Philosophy of the Nyishis', in T. Mibang & S.K. Chaudhuri (ed.), *Understanding Tribal Religion.* New Delhi: Mittal Publications.

Rikam, N.T. 2005. *Emerging Religious Identities of Arunachal Pradesh: A Study of Nyishi Tribe.* New Delhi: Mittal Publications.

Sahu, C., ed. 2006. *Aspects of Tribal Studies.* New Delhi: Sarup & Sons.

Sengupta, S. (ed.). 1996. *Peoples of North-eastern India: Anthropological Perspectives.* New Delhi: Gyan Publishing House.

Showren, T. 2009. *The NYISHI of Arunachal Pradesh: An Ethnohistorical Study.* New Delhi: Regency Publications.

Tara, T.T. 2006. *Nyishi Ancestors.* Guwahati: Bhabani Offset & Imaging Systems Pvt. Ltd.

Tara, T.T. 2008. *Nyishi World* (2nd Edition). Banderdewa: D.B. Printers.

Vidyarthi, L.P. and B.K. Rai. 1985. *The Tribal Culture of India.* New Delhi: Concept Publishing Company.

Online Sources

Barnes, R. H. 1999. 'Marriage by Capture', in *The Journal of the Royal Anthropological Institute,* Vol. 5, No. 1 (Mar., 1999), pp. 57-73. Royal Anthropological Institute of Great Britain and Ireland. http://www.jstor.org/stable/2660963. Accessed on 04/12/2010 01:50.

Bharati, A. 1971. *Anthropological Approaches to the Study of Religion: Ritual and Belief Systems. Biennial Review of Anthropology.* Vol. 7 (1971). 230-282. Stanford University Press. http://www.jstor.org/stable/2949230. Accessed on 22/04/2010 14:46.

Frazer, J. G. 1922. The *Golden Bough: A Study of Magic and Religion* (Abridged Edition). Ebooks@Adelaide. Http://Ebooks.Adelaide.Edu.Au/.

Religion: Encyclopaedia Britannica Online. Accessed on 27 June 2011. http://search.eb.com/eb/article?tocId=9063138.

Rivers, W. H. R. 1914. *Kinship and Social Organisation.* Digitalised by internet archive in 2007. http//www.archive.org/details/kinshipsocialorg00riveuoft.

Xaxa, Virginius. 1999. *'Tribes as Indigenous People of India'. Economic and Political Weekly.* Vol. 34, No. 51. (Dec. 18-24, 1999). 3589-3595.
        http://www.jstor.org/stable/4408738. Accessed on 22/04/2011 14:31.

About the Author

Tame Ramya is currently pursuing his Ph.D. in the Department of Anthropology, Rajiv Gandhi University, Rono Hills, Doimukh, Arunachal Pradesh, India. To his credit, he has presented some research papers on tribes and tribal related issues and has published in national and international journals. He is a member of many professional bodies like IAS, INCAA, LSI, etc.

E-mail: taramya01@gmail.com   Mobile: +91-9402034048.

Chapter 2

The Zeliangrong Movement: Change and Continuity

☆ *R. Longmei*

ABSTRACT

The Zeliangrong are one of the ethnic groups in North-east India who claimed they are a nation and fought against the British raj as well as with other ethnic groups who forcibly encroached upon their lands. In the late 1920s they launched their unification and solidarity movement under Haipou Jadonang. Ever since then, 'nationalism' began and their struggle for the Zeliangrong State within India has witnessed several shifts and changing phases. This chapter generally examines and outlines several shifts and changing phases besides its achievements and failures.

Keywords*: Zeliangrong, Jadonang, Movement, Change, and Continuity.*

2.1 Introduction

Ethnic politics in North-east India have divided the people of the region on ethnic lines. Right from 1947 till date this part of the country has never been peaceful. The diverse ethnic groups located strategically at international borders of Bhutan, China, Myanmar and Bangladesh has seen much violence and bloodshed over the past few decades owing to insurgencies in all North-eastern States with the exception of Arunachal Pradesh and Sikkim. Besides insurgency problems, there are conflicts and confrontations over land use and control, language, identity formation, demographic change, minority and majority relations at the civil level. In recent years there arises an internal conflict among the Zeliangrong within the ethnic politics, such as the *Zeliangrong United Front* (ZUF) in Manipur and *Eastern Nagaland Peoples' Organization* (ENPO) in Nagaland are two stark examples of ethnic politics within the Pan-Naga ethnic politics of North-east. These problems are seen to be continuing and peace is still elusive in India's North-east. With more and more ethnic groups competing in asserting their respective ethnic identities or urging for maintaining their distinct identities life's very depressing. According to Udayon Misra and Girin Phukon, "The politics of the region has been highly ethnicized because each ethnic group demands either sovereign independent state or separate state within the Indian Union on the basis of their respective ethnic identity" (Misra 2000:154; Phukon 2003:1). It may be said that behind this ethnic assertion is the nationalism that awakens the ethno-cultural consciousness. This chapter attempts to trace the origin of the rise of the Zeliangrong nationalism and examine the factors responsible for launching the Zeliangrong movement in the light of change and continuity.

2.2 Zeliangrong

The Zeliangrong refers to the people of three genealogically linked clans – Zeme, Liangmei, and Rongmei (Kabuis and Puimeis) living in a compact and contiguous geographic area that comprises Tamenglong District and a considerable portion of Senapati and Churachandpur Districts of Manipur, Paren District and a considerable part of Dimapur District of Nagaland, and the Halflong Sub-division of the North Cachar Hills District of Assam (now Dimahasao) in India's North-eastern Region numbering around 4.5 lakhs according to 2011 census yet divided by the colonial and independent Indian State boundaries to make them a sort of minorities in those States mentioned. Paul R. Brass clearly mentions in his book *The Politics of India since Independence* about the problems faced by many minority language speakers within the linguistically reorganized states (Brass 1990: 192). *The* Zeliangrong (Zemeis+Liangmeis+Rongmeis, Kabuis+Puimeis) after independence demarcated these geographic regions as theirs and resolved to reconstitute them as one unit under a single administration within the framework of the constitution of India. Perhaps, it was and is for the protection and preservation of their political independence, people, land, culture and customs that the Zeliangrong rallied to their leader's cause. They alleged that unlike in the past they are now living in absolute humiliation as they are made subservient to co-inhabitants of the region who not only dominate them politically but also in other aspects in all the three States mentioned under the Indian constitutional scheme. When they have given their composite tribe name, '*Zeliangrong*', for recognition of their community as Scheduled Tribe by the Indian and State Governments, both the Governments instead of recognizing it rejected it. This shocked the Zeliangrong people. It may be recalled that according to Professor Gangmumei, the term 'Zeliangrong' was coined on 15 February 1947 at the Zeliangrong Conference when the Zeliangrong Council was established at Imphal after they had combined the prefixes of their cognate tribes of Zeme, Liangmei and Rongmei (Gangmumei 2004: 10-11).

Perhaps, the rejection of the Zeliangrong nomenclature has given rise to the unnecessary problem of identity crisis that continues to spoil/threaten the unity of the Zeliangrong people. The Zeliangrong people however despite what others say in pursuant of their historic decision of 1947 began to put in more efforts to achieve their goal of Zeliangrong unity. They began to use the term Zeliangrong when they renamed all the names of their community organizations and their honest struggle for recognizing them as a nation continued. They strongly regarded the rejection (not recognizing Zeliangrong) as a continued attempt to malign, subjugate and divide the Zeliangrong family members. What ever is the case they declared that they would refuse to be divided ethnically and emotionally even though they presently live divided and recognized differently in the three Indian States such as *Kabui Nagas* (Manipur), *Kacha Nagas* (Assam) and *Zeliang* (Nagaland) by the Scheduled Castes and Scheduled Tribes list (Modification) Order of 1956. They argued that according to the Zeliangrong history which is crystal clear to every Zeliangrong there was no such divisions among the Zeliangrong tribes even though cases of internal feuds were inherent in their society. The tribal name of *Kabui Naga* was given by the Meiteis and the British rulers and *Kacha Naga* was given by the British rulers very much to the dismay/disgust of the Zeliangrong people (Pamei 2001:37) they lamented. After independence Indian Government continued to use these tribal names in both the States of Manipur and Assam in addition to the new tribal name called *Zeliang* was given to those Zeliangrong who live in the State of Nagaland. All these tribal names are unacceptable to the Zeliangrong people and hence they wanted the Government of India to recognize Zeliangrong as their composite tribe name to help them live with dignity, respect and recognition. It may be pointed out that like others the Zeliangrong people have rejected these tribal names given to some of their family members by outsiders only after they had attained their national consciousness.

The Zeliangrong people are very proud of their own history which had its long root extending back to pre-British era. Elungkiebe Zeliang wrote that although no authentic written accounts are available as in the case of any other indigenous ethnic tribe of the region regarding their common ancestry, yet the Zeliangrong spirit of oneness was kept alive by their great legends and oral traditions (2010:5). And according to Professor Gangmumei Kabui who is one of the eminent insider writers, the Zeliangrong are undoubtedly one of the earliest migrants who came and settled at Makhel in Manipur from where they moved to Ramtingkebin and then to Makuilongdi. In Makuilongdi the Zeliangrong prospered so greatly that as many as 7777 households were believed to have been lived there. Thus the present Zeliangrong consider Makuilongdi as the cradle of the Zeliangrong culture and custom. It is said that in the course of time however from there they dispersed to their present settlements. The group that moved to the North-west of Makuilongdi came to be known as 'Zeme' (Zemeis) from 'Ramzie' meaning people who live in plain or lowland, the group that moved to the Southern area of Makuilongdi came to be known as Ruangmeis (Rongmeis) and Puimeis from 'Maruang' meaning empty land and the group that lived near Makuilongdi and in its central zone came to be known as Liangmeis (Gangmumei 2004:10-12).

The present Zeliangrong speak five dialects- Northern Zeme, Southern Zeme, Liangmei, Rongmei and Puimei and according to them this is due to their long confinement in their respective geographic locations. They claimed that they have had only one dialect in the past but that dialect is not yet researched out till date by any scholar. In spite of this dialectical problem (lack of common language) the Zeliangrong still claim to have all the essential attributes of a nation – Common Ethnic Origin, Similar Historical Past, Common Linguistic Roots, Common Kinship and Social Structure, and Common Cultural Pattern. They claim to have a rich cultural heritage which is handed down from generations to generations. According to N.K Das, the Zeliangrong had always nurtured the myth of their common ancestry and inherited the rich Zeliangrong culture of the past. Perhaps, this fact and other ethno-cultural similarities among the cognate Zeliangrong brothers seemed to have provided the leaders of the Zeliangrong community to launch a united 'autonomous tribal movement' in the post independent era (Das 1996:20).

The Zemes, Liangmeis, Rongmeis and the Puimeis till today are demanding the authority in New Delhi to recognize Zeliangrong as their collective name for their identity. Earlier there was a problem/crisis of Zeliangrong nomenclature within the Zeliangrong community. The problem was triggered off by the internal differences between the Rongmei and Puimei sections of the Kabui and amongst the pro-Kabui Rongmei and anti-Kabui Rongmei. Many Rongmeis do not think that Rongmei and Puimei are not one and the same though a sizeable section of the Rongmei think so based on the writing of a noted anthropologist T.C. Hodson who used the term 'Kabui' to mean the Rongmei and Puimei. At the same time the Puimei claimed that the word 'Kabui' originated from the word 'Puimei' and hence Puimei should be separated from the Kabui or Rongmei and recognized as Puimei or Inpui. The Puimei strongly urged the Zeliangrong Union in 1959 to recognize Puimei as a separate Zeliangrong cognate tribe by adding the prefix of their tribe's name to Zeliangrong nomenclature. The Zeliangrong leaders after they have pondered long enough decided to convince the Puimeis to accept the Zeliangrong nomenclature 'Zeliangrong' as the new nomenclature 'Zeliangrongpui' would mean mother of Zeliangrong instead of a common tribal name (ZU, 2005:6). It may be recalled that consequent upon a mutual agreement the Puimeis were recognized as a separate Zeliangrong cognate tribe and with that they remained an integral part of the Zeliangrong ethnic group. There is no issue or crisis of Puimei identity after that claimed the Zeliangrong leaders. With regard to the differences of opinion between the pro-Kabui Rongmei and anti-Kabui Rongmei protagonists it may be said that there emerged in the course of time a consensus of opinion among the high majority of the Kabuis or Rongmeis that the term kabui is a Meitei name for the Rongmeis including the Puimeis. This majority opinion has prevented both the pro and anti- Kabui Rongmeis from exchanging tirades against one another (Gangmumei 2004:12). The Zeliangrong are claiming that the creation and introduction of the Zeliangrong nomenclature is the very product of their ethnic consciousness (ethnic identity formation) that had started many years ago. In the backdrop of this, it is therefore pertinent to trace out the roots of the Zeliangrong national consciousness that caused for the rise of the Zeliangrong.

Perhaps, questions as to when, why and how the Zeliangrong considered them as one people and started their political movement are of immense significance as one wants to investigate the reason or reasons for their rise and fall and then rise. As pointed out earlier there is no exact record of these people prior to the British occupation of their country. If one has to rely on their legends and oral traditions it is rather easier to conclude that these people lived as one big family (a life independent of others) and had on and off relationships with other ethnic groups. According to them (their legends and oral traditions) though internal contradictions were there yet they have not encroached upon other people's land or declared wars against any ethnic group except only on those who had encroached upon their land and declared wars against them. The problem started for them sooner the British led by Francis Jenkins and R.B. Pemberton penetrated into their country through northern Zeliangrong in 1832. In 1863 a Zeliangrong village, Asalu in North Cachar Hills, Assam was turned into the first British political center for the entire Naga country wrote Piketo Sema (Sema 1992:8-10). Since then the entire Zeliangrong areas were targeted by the British and their allied forces. The Zeliangrong seemed to have reacted sharply and became more defensive than offensive without their leaders at that point of time.

Then came Haipou Jadonang of Kambiron and took up the task of proclaiming himself as a King sent by God (the much expected Messiah King about whom for a long time there had been a prophecy in the Zeliangrong country who would drive out the British and rule over all the Naga tribes). Perhaps, with him and under him the story of the Zeliangrong got changed. The Zeliangrong believed that Jadonang was really the one sent by god and so they followed his instructions. They offered him tributes not because they fear him but out of much reverence and solidarity of the entire Zeliangrong people as one community. The Zeliangrong were now turned more offensive sooner they launched their political movement under his leadership and his confidant Rani Gaidinliu. It seems that Haipou Jadonang who had identified several menaces to the Zeliangrong people that are responsible for their rise could not simply forget the terrific and horrific Kuki infliction on the Nagas which have resulted killing of many innocent Zeliangrong and Tangkhuls during the Kuki Rebellion (1917-20) partly due to the British inactiveness (insensitive) to protect them (Pamei 1996:36).

2.3 Zeliangrong Identity Formation and the Political Movement

Haipou Jadonang (1905-1931) knew that the Zeme, Liangmei and Rongmei tribes of the then Naga Hills, Manipur and North Cachar Hills of Assam were not united even though they had always nurtured the myth of their common ancestry. The people of these three cognate tribes were yet to secure their political integration so that they could consider themselves part and parcel of a single nation and owe their loyalty to a single centre of power. Haipou Jadonang also had in mind the establishment of the Naga Raj by uniting all the different Naga tribes inhabiting the then Naga Hills under his rule and no doubt their integration into one nation-state involves the problem of nation-building (developing a sense of national identity as a rallying point for the people, cutting across their group loyalties based on religion, race, caste, language, region, language, culture, etc.) and state-building (respect for authority and for the prevalent method of rule) which was not going to be easy. But he was very much determined to introduce his ideas into actions. So with that strong determination and a clear vision Jadonang had launched an ethnic movement in the late 1920s. Many scholars have taken the Jadonang Movement to be purely an anti-British revolt but a closer look at it reveals the following objectives:

(a) To revitalize and reform animistic tribal religious rites and beliefs, in the face of advancing Christianity (John Parratt 2005:44);

(b) To unite the three cognate Zeliangrong tribes (three ethno-culturally allied tribes Zeme, Liangmei and Rongmei) into one nation (Gangmumei 1991:135);

(c) To take vengeance on the Kukis who had forcibly encroached upon their areas and inflicted the Zeliangrong people and the Tangkhul Nagas during the Kuki Rebellion (1917-20) (Lal Dena 1991:131);

(d) To revolt against the British rule and to drive them out of the Zeliangrong areas (S.K. Barpujari 2003:247); and

(e) To establish a Naga Raj or 'Makam Guangdih' (Gangmumei 1991:135).

But the first Zeliangrong ethnic assertion and crystallization movement under Jadonang was short-lived owing to the swift punitive actions of the British government. Sooner the Jadonang movement was launched and before it could effect havoc the British recognized Jadonang as their enemy number one and sought all possible helps from other warring/hostile tribes to execute him at the earliest. Here the roles played by the Kukis, Meiteis, and some of his own people were so crucial for the British to successfully counter his movement. So with their valuable helps on 19 February, 1931 Jadonang was arrested and confined him at Imphal jail till he was finally executed by hanging on 29 August, 1931 at 6 a.m. on the bank of river Nambul behind the jail. The then Political Agent of Manipur Mr. Higgins and Government of India had secret consultations on the execution of Jadonang even though they knew that Jadonang was nothing to do with the dead of four Manipuri traders in his areas. Jadonang, a gifted leader and the brave freedom fighter was simply made the scapegoat (Raghavaiah 1971:240-245). His dream could not be fulfilled and his movement had been nipped in the bud but the fire of his movement did not die down. The execution of Jadonang who was considered as the messiah king of the Zeliangrong people greatly shocked the Zeliangrong people and at the same time they were infuriated at the way Jadonang was executed (Zeliang 2005:22). Thus filled with anger they continued the Jadonang movement in a more resurgent manner under his faithful confidant and follower Rani Gaidinliu (Barpujari 2003:247). Many Nagas reviled the Zeliangrong leaders for continuing the Zeliangrong Movement but according to the author of *The Trial from Makuilongdi: The Continuing Saga of the Zeliangrong People* Namthiubuiyang Pamei the success of the Zeliangrong will enhance, not weaken the health and strength of Naga family (Pamei 2001: vii). Neketu Iralu, one of the greatest contemporary Naga intellectuals agreed with this view when he wrote the foreword of the author's book mentioned.

2.4 Rani Gaidinliu

Rani Gaidinliu (1915-1993) raised a new standard of rebellion against the British Government from Manipur, Naga Hills and the North Cachar Hills of Assam. During her leadership the Zeliangrong movement came down heavily on the British rule and so the British government was equally determined to crush the movement. As in her mentor's case Gaidinliu's efforts for the realization of Jadonang's dreams soon invited the British wrath. She was forced to go underground to direct the rebellion not so long after she declared herself as the next Zeliangrong leader. Searches for Gaidinliu were instituted and President of the then Manipur State Durbar offered a cash reward for the arrest of Gaidinliu and any village giving information about her was promised a remission of ten years' taxes. In its efforts to nab Gaidinliu many of her strong supporters/followers were meted out with serious punishments. In spite of this Gaidinliu succeeded in connecting the Zeliangrong Movement with the mainstream Indian National Movement for freedom. Mahatma Gandhi and others recognized her and her followers as brave freedom fighters. But before Gaidinliu and prominent Indian leaders could unite together to forge a united front against the common enemy in this part of the country Gaidinliu was arrested by Captain Macdonald in 1932 and her movement was more or less suppressed after that. When she was brought to Imphal, Higgins sentenced her to life imprisonment. Pandit Jawaharlal Nehru learnt about Gaidinliu's extraordinary bravery and her movement when he paid a visit to Assam in 1937. He was impressed by her activities and so when he spoke in a press conference he described her as the Rani of the Nagas. Nehru then tried his best but succeeded only in 1947 to get her released from jail to retire from active political life (Gangmumei 2002:72).

It may be recalled that the seed for ethnic unity sown by Jadonang and Rani Gaidinliu got germinated when the Kabui Samiti was formed by the three ethno-culturally allied tribes- Zemei, Liangmei and Rongmei on 7 March 1934. The predecessors of the Kabui Samiti, the Agangmei (1925) and the Ching-Zang (1927) were organized by the valley Kabuis and later joined by the Kabuis or Rongmeis of the Manipur hills and the Kacha Nagas (Zemeis and Liangmeis) of the then Cachar and Naga Hills. Perhaps, the very idea for the formation of this stronger union of the Zeliangrong people was given to them by the then SDO C.S. Booth after he had a discussion with the Zeliangrong leaders regarding their origin and common ancestry. On 1st April 1934 the representatives of the three Zeliangrong cognate tribes once again gathered at Tamenglong under the banner of the Kabui Samiti and performed the customary ritual/rite called Chuk-Thoibe/Chuk-Thoibou/Chuk-Sumei and they declared that day as the day of Reconciliation or the day of Zeliangrong Solidarity in the history of the Zeliangrong community (*Souvenir* of ZU (AMN)). The Kabui Samiti worked for the interests of these allied tribes till it gets transformed into Zeliangrong Council on 15 February 1947. In this year the representatives of the three Zeliangrong cognate tribes from Assam, Manipur and Nagaland met for the first time since the defunct of their earlier Kabui Samiti at Keishamthong and under the chairmanship of L. Lungalung, formed the *Zeliangrong Council* and coined the term 'Zeliangrong', by putting together the prefixes of their tribe names after emphasizing their close relationship/ common ancestry. The main aim of the Zeliangrong Council was to work for the economic, social, educational and political advancement of the Zeliangrong.

Unfortunately, the Zeliangrong Council was left defunct since its inception and in its place the Manipur Zeliangrong Union functioned for the unity and solidarity of the Zeliangrong people. Since February 1960 the word Manipur was replaced by Zeliangrong and it was the Zeliangrong Naga Union that had took up the burden of continuing the socio-political movement of the Zeliangrong people. In 1961 the defunct Zeliangrong Council tried to revive its organization and resume its functions in its meeting held at Majorkhul but many of its pivotal figures instead of joining it either joined the mainstream Indian politics or Naga politics leaving the Zeliangrong Council to die. In 1964 the three Zeliangrong bodies – Zeliangrong Council, Zeliangrong Naga Union and Rongmei Association held a joint meeting at Heningkunglwa where a new decision was taken. The new decision was that of entrusting the Zeliangrong Naga Union to continue to spearhead the Zeliangrong movement for recognition of the Zeliangrong people as a nation instead of the Zeliangrong Council. Lungbe, the President of the Zeliangrong Council was present in that meeting but he did not raise any objection but pledged to render the needed support to the Zeliangrong Naga Union in its manoeuvre. Then on 11 October 1980 the three Zeliangrong bodies – *Zeliangrong Naga Union, Zeliangrong Council* and *Rongmei Association* formed the *Zeliangrong People's Convention* (ZPC).

With the formation of the ZPC the Zeliangrong movement/politics has reached its peak and entered a crucial phase. The two preceding phases – Zeliangrong Identity Formation phase (since Jadonang Movement till 15 February 1947) and Zeliangrong Nomenclature Movement phase (since 15 February 1947 till the early 1960s) laid the foundation for this crucial phase. The main goal of the ZPC movement was creation of a separate Zeliangrong homeland within India and according to ZPC leaders the decision taken on 30 November 1981 (Gangmumei, 2004: 234) to demand for a separate homeland for the Zeliangrong people within India was based on the aged old Zeliangrong history. They declared that the Zeliangrong are a nation and India is ignoring this knowingly. They also said that to understand the nature of the ongoing struggle of the Zeliangrong people in its correct perspective it is pertinent to recall and recollect the political events that shaped the Zeliangrong history thus far which, according to them are being reproduced in these points (Some of these points formed the main bodies of the two memorandums submitted to the Indian Prime Ministers by the Zeliangrong leaders one on 14 October 1964 and the other on 26 April 1982):

1. That the Zeliangrong people are one of the earliest migrants despite of the fact their origin is still shrouded in mystery till date as in the case of any other migrant communities in the India's North-east.

2. That the Zeliangrong were peace loving race who lived a life independent of others. They lived as one big family and had on and off relationships with other ethnic groups prior to the British occupation of their country. According to them (their legends and oral traditions) though internal contradictions/feuds were there yet they have not encroached upon other people's land or declared wars against any ethnic group except only on those who had encroached upon their land and declared wars against them. The problem started for them sooner the British led by Francis Jenkins and R.B. Pemberton penetrated into their country through northern Zeliangrong in 1832 and made one of their ancestral villages Asalu in North Cachar Hills, Assam, the first British political center for the entire

Naga country in 1863. Since then the entire Zeliangrong areas were targeted/attacked by the British and their allied forces. The Zeliangrong people have suffered much at the hands of the British and also at the hands of their buffer states. During the Kuki Rebellion the British not only failed to protect the Zeliangrong people upon whom they had imposed heavy household taxes. The Kuki rebels in the name of fighting against the British rule ransacked many Zeliangrong and Tangkhul Naga villages and took away many precious lives. Thus, the Zeliangrong Naga led by Jadonang and Rani Gaidinliu revolted against the British rule and those who have inflicted upon them in the past. After independence Pt. Jawaharlal Nehru recognized the Jadonang and Rani Gaidinliu as freedom fighters and respected their movement but the genuine voices/cries of the Zeliangrong peoples for recognizing them as a nation and for organizing their entire contiguous areas into one political unit in independent India was not at all heard by the new leadership. The leaders of the Zeliangrong community continued to believe in the Indian sense of justice and fairplay. So they launched their Nomenclature Movement under leadership of Rani Gaidinliu for recognition of their composite tribal name Zeliangrong since 1947 under the leadership of Rani Gaidinliu for they think that is the first step towards their ultimate political goal. But their voices have been fallen into the deaf ears and their efforts went in vain.

3. That in 1960 Rani Gaidinliu was forced to go underground to continue and direct the Zeliangrong movement. In October 1964 she met K. Kalanlung, leader of the Zeliangrong Regional Council who had submitted a memorandum to former Prime Minister Lal Bahadur Shastri on October 14 1964 in which he raised the issue of Zeliangrong Political Integration. When Rani Gaidinliu came overground in 1966 both India and her people welcomed her. But she in no time went to Delhi and met former Prime Minister Indira Gandhi reiterating the issue of Zeliangrong political integration and demanding a Zeliangrong district comprising of the present Tamenglong, Zeliangrong ancestral areas of Assam and Nagaland. Mrs. Gandhi after a patient hearing expressed her opinion in favour of planned development of Zeliangrong region or areas in stead of granting what Rani Gaidinliu wanted for her people. Then Rani Gaidinliu wrote to Mrs. Gandhi again on 9 March 1966 to reconsider her political demands.

4. That after the merging of the United Naga Integration Council (UNIC) that demanded Manipur Nagas' integration with rest of the Naga people either in independent or greater Nagaland within India with the Indian National Congress (INC) on 4 August 1972 after the statehood was granted to Manipur Rani Gaidinliu once again pressed the former Prime Minister of India for granting the demand for Zeliangrong integration. The Zeliangrong organizations such as the Zeliangrong Naga Union and Association also submitted memorandums in support of the demands put forward by Rani Gaidinliu their leader. Then on 1 February 1974 yet again Rani Gaidinliu wrote to Mrs. Gandhi who in her reply for the first told Rani to have a mass

contact with her people in the Zeliangrong areas and assured her that she would examine the question further and give a serious thought to it to see to possibility of considering the demands of the Zeliangrong.

5. That in 1975 and 1976 after the Mid-Term Election of 1974 in Manipur the Zeliangrong leaders convened the first and second Zeliangrong Conferences on Zeliangrong political integration. Rani Gaidinliu had already toured many Zeliangrong areas for the purpose. The third conference held in 1980 was the most significant one in the history of the Zeliangrong people because it was in this conference that the Zeliangrong People's Convention (ZPC) was formed with Rani Gaidinliu as its president to demand for a separate Zeliangrong homeland within the Indian Union that affected the on-going mainstream Naga politics and their immediate neighbors of the region. The ZPC had four crucial meetings first in November 1980, second in 1981, third in June 1981 and the fourth in November 1981. The outcome of these meeting was the decision to demand for Zeliangrong homeland which was turned down by the former Prime Minister Indira Gandhi saying that it was not feasible and no longer possible after reorganization of states. She said that had the Zeliangrong people raised their voice for their homeland in 1954 India could have easily granted the same that time. Rani appealed to her and vowed afresh before her that she would not stop till the goal or her mission was achieved.

6. That the leaders of the ZPC presented their case before former Prime Minister Rajiv Gandhi still believing in him that their right, democratic and genuine political demands would be accepted and considered. But like his mother Rajiv Gandhi was interested only in giving anything short of homeland. For carrying out a planned development programme he directed the North-eastern Council (NEC) team to tour all over the Zeliangrong areas in 1985 to see the ground realities and to report back. Several rounds of talks were then held between the ZPC leaders and the Government of India at the Prime Minister as well as Secretary levels from 1986 to 1992 on the Zeliangrong issue.

7. That on 17 February 1993 Rani Gaidinliu passed away. On 23 February 1993 State funeral was given to her with all respects and all the ZPC leaders were present. On 15 August 1993 after a second thought the ZPC movement was halted by its working president.

It is not clear why the ZPC movement was abruptly suspended. But many opine that several factors must have influenced the ZPC leaders to halt the ZPC movement. Some of these factors could be the weaknesses of the ZPC leaders themselves of the time and the problem of mass support for its cause or opposition to ZPC politics or demands instead of cooperation from the then Zeliangrong politicians. Stiff opposition from the neighboring tribes and the concerned State governments to the demands of the ZPC could be another factor. But many do not rule out the possibility of interference by the Naga as well as the Manipur valley insurgents against the larger interests of the Zeliangrong people. The prolonged sickness of Rani Gaidinliu that rendered her inability to lead and guide the ZPC movement and the early 1990s Kuki-Naga ethnic clashes that forced both the Naga and Kuki leaders busy could have adversely affected the ZPC movement. Perhaps these are some amongst others that could have played their roles in forcing the ZPC leaders to put a halt to their movement. The ZPC leaders have never declared the reason or reasons for the suspension of their ZPC movement. A larger section of the Zeliangrong people still believes even today that disunity among the Zeliangrong leaders could be the most important factor behind the collapse of the Zeliangrong genuine political movement.

Two years later that is on 25 November 1995 at New Jalukie (Nagaland) the name of the Zeliangrong Naga Union was changed to Zeliangrong Union (Assam, Manipur & Nagaland). This Zeliangrong organization has worked for the good of the whole Zeliangrong community till it got split into Zeliangrong Union (AMN) and Zeliangrong Baudi (AMN) in 2010. Even today these two Zeliangrong bodies are functioning separately because both of them claim to be the apex Zeliangrong body of the Zeliangrong people. But both of them agreed to have the Zeliangrong Interim Body (ZIB) till agreement between the two is arrived at that will represent Zeliangrong people anywhere in the world.

In 2005 some of the leaders of Zeliangrong tried to reassert Zeliangrong identity and restart the Zeliangrong Homeland Movement. They organized a meeting at Tamenglong on 1 April 2005 and the existence of the Zeliangrong state/country was explicitly expressed by placing lighted candles collectively on the displayed map of Zeliangrong Region. But the Rongmeis (Kabuis) and Puimeis that formed bulk of the said ethnic group did not take part in it. This shows that the Zeliangrong people still lack their national unity and are still lagging behind others in this regard (Singh 2005:27-28). Perhaps this lack of unity among the Zeliangrong continues to be the stumbling block on their way to self-determination.

The Zeliangrong people are yet to settle their differences and live in unity. When some Zeliangrong formed the Zeliangrong United Front (ZUF), a militant group in 2011 and restarted the halted Zeliangrong/ZPC movement not all the Zeliangrong are in favour of it and this is not surprising enough. How can it be supported by all Zeliangrong when the Zeliangrong are still struggling to live in unity? Definitely the Zeliangrong are divided. While some Zeliangrong say ZUF is not required keeping in mind the need to back the on-going Indo-Naga Peace Process that is trying to hammer out an honourable political solution to the vexed Naga problem/issue others are feeling the need to back the ZUF movement keeping in mind the division in Naga society and the snail-slow Indo-Naga peace process and the Zeliangrong history.

2.5 Zeliangrong and the Naga Political Movement

It may be recalled that in 1957 the Zeliangrong people joined the mainstream Naga Political Movement and the Federal Government of Nagaland (FGN) led by A.Z. Phizo who recognized the Zeliangrong Region as the seventh province of sovereign independent Nagaland. But interestingly the Zeliangrong Movement had not yet come to an end as expected by India after the Zeliangrong people raised the issue of Naga political integration. Perhaps since 1957 till the merging of the UNIC with the INC in August 1972 the Zeliangrong people submerged themselves deeply into the Naga politics. They joined hands with the Naga national leaders both in their underground as well as overground activities and even sacrificed many of their precious lives for the greater Naga cause. In Manipur, first, they openly supported the demand of the United Naga Integration Council that was the integration of the Manipur Nagas with the rest of the Naga people either in independent or greater Nagaland. Then since the inception of the United Naga Council (UNC) they have been part and parcel of UNC that works for the Manipur Naga interests as apex Naga body in Manipur. Since the beginning of the current Indo-Naga peace process the UNC is deeply involving in supporting/backing the Naga collective leadership for finding out an honorable solution to the vexed Naga issue.

According to the Zeliangrong leaders the Zeliangrong Naga have enough land and resources and people to form a separate sovereign independent nation outside the Indian Union or a separate state within the Indian Union. To live together was their inherent right and the political integration of the Zeliangrong was their sole aim. But since the Naga political movement was aimed at integration of all the Naga areas they decided to halt their political movement keeping in mind the Greater Naga interest or cause. As mentioned earlier some of the Zeliangrong are so impatient that they formed a militant organization called *Zeliangrong United Front* (ZUF) with the objective of continuing the halted Zeliangrong movement. According to the Zeliangrong People's Memorandum of Understanding (MoU) signed on 28 March 1996, the Zeliangrong people however declared that they are an active partners in the present struggle for Naga Sovereignty. Despite differences of opinions between them and the Naga national workers they hold on to the aspirations of the Naga people together. They declared that they are committed to the greater Naga cause for unity, peace, understanding and cooperation. Thus, in pursuant of this memorandum the present Zeliangrong are backing the Naga national leaders instead of encouraging or cooperating divisive forces in the Naga inhabited areas (Pamei 2006:148-149). N. Pamei (2006) has discussed in great detail about the involvement of the Zeliangrong people in the Naga political movement in his book, *The Trail from Makuilongdi: The Continuing Saga of the Zeliangrong People*.

2.6 Conclusion

In conclusion, the Zeliangrong movement started by Jadonang was continued by Rani Gaidinliu was purely an anti-British and anti-Kuki when it was launched. This Zeliangrong struggle may be taken as the phase of Zeliangrong Identity Formation. In the post independent period this movement got transformed into a Zeliangrong Nomenclature Movement till it was fully transformed into a full-fledged Zeliangrong Homeland Movement under the banner of ZPC using the unique Zeliangrong ethnicity and history against the Indian Union. According to contemporary Zeliangrong leaders, the Zeliangrong movement is not yet fizzled out; it is just halted... if the Naga political solutions do not give enough political benefits they (Zeliangrong) movement would begin again in a much formidable form for the achievement of all the set forth objectives right from the beginning whether in present or future Nagaland. As of now, despite of rift emerging in Zeliangrong society, the

Zeliangrong Union (AMN) and Zeliangrong Baudi (AMN) are actively engaged in all developmental activities in the Zeliangrong region and are giving all possible supports to *United Naga Council* (UNC) the apex Naga body in Manipur that supports the on-going Indo-Naga peace process. The Meitei and Kuki organizations in Manipur have however ridiculed the political stance of the UNC as it could have adverse effects both of them (AMUCO, 2004).

The Meitei struggles for Manipur Integrity and Unity of the people of Manipur since 2001 while the Kukis demands for a separate Kuki homeland or state that have already launched their counter-homeland movement in Manipur affecting both the Nagas and the Meiteis in Manipur. According to *National Socialist Council of Nagaland*, the Kukis were settled in the Naga territories by the British and the Meiteis who now launch for homeland movement for the snatched portions from the Naga territories (NSCN 2005:3). The *Kuki Inpi Manipur* (KIM) is yet to come in terms with the Naga militant groups that get involved during the Kuki-Naga ethnic clashes from 1990-1998 and killed many Kukis as alleged by them (KIM, 2004:3; KMHR 2004:5). On the other side, impatient with the snail-slow progress of the Indo-Naga peace process and owing to culmination of numerous issues and factors a section of the Zeliangrong society formed the *Zeliangrong United Front* (ZUF) in 2011 with the primary objective to fully revive the halted Zeliangrong Movement and to realize all the unfilled dreams of the Zeliangrong people. According to one of its leaders, protection of the Zeliangrong region (land and people) is the core objective of the ZUF. He said that the ZUF is in full control over the Zeliangrong region since its inception where only the Zeliangrong people shall control, and the NSCN supremacy over the Zeliangrong people and their land have been destroyed (*The Imphal Free Press, March 7, 2011*). But ironically it is not yet clear till date whether this ZUF is fully backed by the whole Zeliangrong community or not, only time can tell us. Whatever, the fact is that a section of the Zeliangrong people have once again began their march towards achieving the declared Zeliangrong political goal by Haipou Jadonang when the Naga as a whole has entered into a crucial phase of their political journey (ZSUG 2009:33). However, it would be wrong to say that with the existence of ZUF the Zeliangrong general public are no longer backing the on-going Indo-Naga peace process and are encouraging a divisive force/ element within and outside the Zeliangrong region. The crux of the matter is that the Zeliangrong are the very blood of the UNC that renders all possible supports to the NSCN-IM that is negotiating for the Naga political solution and there is no denying or doubting about that being Naga.

References

Barpujari, S.K. 2003. *The Nagas: The evolution of their History and Administration (1832-1939).* New Delhi: Spectrum Publications, p. 247.

Das, N. K. 1996. 'Cultural Identity and Tribal Heritage of North-east India' in Kalyan Kumar Chakravarty (ed.). *Tribal Identity in India: Extinction or Adaptation*.Bhopal: Indira Gandhi Rashtriya Manav Sangrahalaya, p. 20.

Dena, Lal (ed.). 1991. *History of Modern Manipur (1826 – 1949).* New Delhi: Orbit Publishers – Distributors, p. 131.

Haokip, P.S. 2008. *Zale'ngam – The Kuki Nation.* KNO Publications, (for private circulations), p.478.

Kuki Inpi Publicity Wing. 2004. *Brief Outlines of the Kuki Inpi (Kuki Government).*

KMHR. 2004. *The Indigenous Kukis and Fellow Indigenous Peoples between Former India and Burma.*

Kabui, Gangmumei. 1991. 'Zeliangrong Movement under Jadonang and Rani Gaidinliu (1930-1949)' in Lal Dena. *History of Modern Manipur (1826-1949).* New Delhi: Orbit Publishers-Distributors, p. 135.

Kabui, Gangmumei. 2004. *A History of the Zeliangrong Nagas: From Makhel to Rani Gaidinliu.* New Delhi: Spectrum Publications, p. 10-11.

Kamei, Gangmumei (ed.). 2002. *Jadonang: A Mystic Naga Rebel.* Imphal: Published by the Author, p. 72.

Misra, Udayon. 2000. *The Periphery Strikes Back: Challenges to the Nation-State in Assam and Nagaland.* Shimla: Indian Institute of Advanced Study, p. 154.

NSCN Chairman's Address. 2005. Published by *Ministry of Information & Publicity, GPRN* Pamei, Ramkhun. 1996. *The Zeliangrong Nagas: A Study of Tribal Christianity.* New Delhi: Uppal Publishing House, p. 36.

Phukon, Girin. 2003. *Ethnicisation of Politics of North-east India.* New Delhi: South Asian Publishers, p.1.

Paul, R. Brass. 1990. *The Politics of India since Independence.* Cambridge: Cambridge University Press, p. 192.

Pamei, Namthiubuiyang. 2001. *The Trial from Makuilongdi: The Continuing Saga of the Zeliangrong People.* Tamenglong: Gironta Charitable Foundation, p. vii (Foreword by Neketu Iralu).

Pamei, Namthiubuiyang. 2006. *Naga Crucible: 50 years of a Movement in Search of its People.* Tamenglong: Gironta Charitable Foundation, pp. 148-149.

Parratt, John. 2005. *Wounded Land: Politics and Identity in Modern Manipur.* New Delhi: Mittal Publications, p. 44.

Raghavaiah, V. 1971. *Tribal Revolts.* Andhra Rashtra Adimajati Sevak Sangh: Nellore, pp. 240-245.

Sema, Piketo. 1992. *British Policy and Administration in Nagaland 1881-1947.* New Delhi: Scholar Publishing House, pp. 8-10.

Singh, Aheibam Koireng. 2005. 'Some Reflections on Land Questions and Ethnic Crisis in Manipur', presented at the MAKKAIAS sponsored National Seminar on *Land Problems and Ethnic Crisis in* North-east *India,* organized by Department of History, Manipur University from 27-28 July.

Singh, Aheibam Koireng. 2006. *The Kuki-Naga conflict in Manipur (1990-2000).* Unpublished PhD Thesis on Department of Political Science, Manipur University, Imphal, p. 65.

Souvenir. 2004. *All Manipur United Clubs Organization*, 7th Manipur Integrity Day, 4 August.

Souvenir. 2005. 'Crossroad'. Published by the Zeliangrong Union (AMN), 71st Zeliangrong Solidarity Day Celebration at Tamenglong under the theme '*Unity is Strength*', 1st April . The Imphal Free Press, March 7, 2011.

Unison. 2009. Zeliangrong Students' Union. Guwahati (ZSUG), Vol.1.

Zeliang, Elungkiebe. 2010. 'Rani Gaidinliu: A Challenge for the Development of Women Leadership in North-east India' in Lungsanliu & Rampaukopoing Michui (eds.). *Perspectives: Reflections on Issues Challenging Zeliangrong Church and Society.* Dimapur: Zeliangrong Theological Forum, p.5.

Zeliang, Thunbui. 2005. *Haipou Jadonang (1905-1931).* Guwahati: Heritage Foundation.

About the Author

R. Longmei is a Research Scholar in the Department of Political Science, North-eastern Hill University, Shillong, Meghalaya-793022.

Email: rimmei.longmei@gmail.com;   Mobile Number: 9612729397.

Chapter 3
The Ethno-history of Kom-Rem of Manipur

☆ *Doupu Kom*

3.1 Introduction

North-east India refers to the eastern most region of India, composing the states of Arunachal Pradesh, Assam, Manipur, Meghalaya, Mizoram, Nagaland, Tripura and Sikkim.[1] It is ethnically, linguistically and culturally very distinct from the rest of India.[2] It is a home of various ethnic people groups.[3] There are thirty two scheduled tribes in Manipur and the *Kom-Rem* tribe is one among them. Most of the Kom-Rem dwells at the foothills of the Manipur valley and receive production of harvest, both from the hills and the valley. From ancient time, the Kom-Rem has had their own distinct religious-cultural heritage which binds them together in their community. They have no clear cut distinction between secular (cultural life) and sacred (religious life). They practice a series of rites and ceremonies in their individual and corporate life. Thus, this chapter highlight the ethnic roots of the people and a profile of the socio-religious and traditional heritage of the Kom-Rem.

3.2 The Ethnic Roots of the People

The Kom-Rem is one of the indigenous tribes of the Indian union. They belong to the Mongoloid racial stock under Chin-Kuki-Mizo speaking groups.[4] "They exhibit a short stature, mesocephalic head shape, leptoprosopic face and a high frequency of the mesorrhine nasal index".[5] An etymological study of the word '*Kom*' gives an insight into the origin of the community bearing the name. There are two theories regarding the word root of '*Kom*'. Firstly, the name '*Kom*' was given to those who had the culture of wearing a '*lukom*'(turban). This concept is derived from two words *i.e.* '*lu*', means 'head' and '*kom*', means 'wrap' or 'bind', the Kom-Rem are a community who had the culture of keeping their hair long and cared for them with *sahrik* (oil obtained from the pigs fat). The hair could be wrapped in a turban (*lukom*), and both men and women used to wear in the same style.[6] Another theory regarding the word '*Kom*' was that, the community in their wandering and settlements had come across Meitei people who asked them about their origin. In answering their question, the Kom-Rems said that they had come from caves.[7] The Meitei literally translated the word '*Kom*' means 'caves' or 'great hole' in their own language.[8] During nomads, they came across a way between two narrow hills, cave like structure, and thus they came to be known as '*Kom*' people. Even their traditional song proves it, which is as follows: *"Kan hong sok eh, Kan hongs ok eh, khur-pui ah kan hong sok eh".*[9] This means "we came out, we came out, from the cave, and we came out."

Even for the word '*Rem*', there are two interpretations. Firstly it simply means 'people'. So, Kom-Rem means 'cavemen'.[10] According to another interpretation, it means 'people who eat meat'.[11] Mythological stories and traditional song, record that when the Kom-Rem people emerged out of a cave, the mouth of which was guarded by a tiger, the people in order to escape wore striped cloth which resembled the tiger's skin. They came out and made friendship with the tiger and killed him. Out of joy, everyone drank wine and ate the flesh of the tiger and rejoiced.[12] Such people who ate the tiger's meat called themselves *Rem*. The following traditional song proves it:

> *"Korkei ranging raloi amanram, Ami kanthem e amtin kanser a kan amtin ser korkei amanram."*

This means, "The tiger has killed us but we are expert in making cloth and our cloth leads to the end of the tiger". Generally, the Kom-Rem people are slim and dark and are on an average, shorter than the Naga tribes. Girls are lovely and attractive in their facial structure, graceful and charming in appearance.[13]

3.3 Migration and Settlement

It is generally considered and accepted that the Kom-Rem people belong to the Kuki-Chin race, sub-family of the Tibeto-Burman or Indo-China family.[14] Though their original homeland is unknown, there are oral traditions which have no documental proof which tell us about their origin. Between 100 B.C. to 100 A.D., the Kom-Rem were believed to have entered Burma moving towards the Arakan Hills and Bay of Bengal moving to Tripura under their Chief Nei-thot-hla. They were said to have been driven away from the land and from the entire range of Chin Hills by a stronger tribe known as Soktes, residing in the northern parts of Chin Hills in Burma.[15] These are all oral traditions passed from generation to generation and there is no written record.

The Kom-Rem people came to Manipur from the hills south of Manipur in Burma in the sixteenth century during the reign of Maharaj Gambhir Singh.[16] The Kom-Rem people are found scattered in all parts of Manipur forming small groups in villages.

It is believed that the oppression by the superior and stronger tribes in Manipur caused this dispersion into all parts of Manipur. Other reasons given are for agriculture prospects and to enjoy the fruits of the hills and plains. Whatever may have been the reason, the Kom-Rem people are found in the district of Churachandpur and Sadar Hills inhabiting the foothills of Manipur valley surrounded by the other tribes like Kuki-Chin in the hills and the Meiteis in the plains, thus being influenced by them which can be seen in the slight differences in language and customs among the Kom-Rem people settling in the different parts of Manipur.[17]

3.4 Social Life

The Koms have their own distinct social life. Their society is a classless society. There is no discrimination among the people on the basis of class. Every person is treated equally. The rich will help the poor and the poor will honor the rich. The social life of Koms is marked with mutual love and harmony.[18]

The Kom-Rem people have a distinct social life of their own and their society was and is still now patriarchal in character. Each family bears the clan name of the father, who stands as the head of its family with authority in all matters of family dealings. In earlier times, the bond of love between brothers was very strong that everything in the family, particularly family properties belonged to them in common. Later on, this concept was changed into new one, in which the properties either go to the eldest son or the youngest son. Thus, this custom has become quite flexible today.

Among the Kom-Rem people marriage within the clan was not restricted, though marriage among close blood relations was strictly prohibited. The practice of both 'love' and 'arranged' marriages are found in the Kom-Rem society though a man is expected to marry a girl from his maternal uncle's clan.[19] A girl usually leaves her father's house after marriage. In the earlier times, the bride-groom's parents pay the customary prescribed bride-price to the parents of the bride, which is not strictly followed at present due to the changing conditions of the social life of the people. As per the customary laws, the eldest in the family should marry first and then the younger. Otherwise, if the younger ones get married first, then there are many customary laws that command the younger to pay a sum to his elder, which is called as *U-Khe*.[20] Divorce is uncommon except in the case of adultery committed by either of the partners. In such case, the one who is caught in adultery has to pay a fine that is known as *Hmaimuk*, of one mithun. In the case of a husband who dies leaving a young wife, she can leave her children, if any, with her in-laws and return to her father's house and is allowed to remarry which is called *Inhlam-Inle* (returned home).[21] Kom-Rem even had a centre or institution of learning and discipline where the boys were trained in etiquette, obedience to parents and good behavior before they attain puberty. This practice of system is known as *Sawm* (bachelor's dormitory).[22]

3.5 Economic Life

The Kom-Rem people were self-supporting and self-sufficient in their economic life and did not depend on people outside the village for their livelihood. The hindrance for their economic development for a long time was because of ignoring the development of the outside world. But this situation has changed now. Like other tribes in the country, the economic position of the Kom-Rem tribes is very insecure. Trade is very limited. They are very ignorant in the field of business and in the economic policy. They are so innocent and sincere that they are easily cheated by people. They always get less amount of money in comparison with the money spent for their production.[23] The Kom-Rem enjoyed both the fruits of the hills and of the plains as majority of them lived at the foot of the hills. Agriculture is the main occupation of the Kom-Rem and rice is their staple food. They depend entirely on the monsoon rain for agriculture. Besides this, the people of Kom-Rem are also engaged in other economic activities like rearing pigs, gardening, cattle farming, poultry, etc. The barter system also was practiced in the society. There was also Jhum cultivation in the hills where varieties of plants were grown.

The women-folk were known for their skill in weaving cloth with beautiful patterns on them and provide sufficient clothing for the whole family. The traditional cloth woven for men is called *Pase Pon* and the clothes of the ladies were called as *Khamtlang* and *Ponlak*. These clothes were woven from the design taken from the stripes of the tiger and the python.

The majority of the Kom-Rem people were poor. In spite of their hard work, they do not have enough food to eat and clothes to cover themselves. They need to work hard in order to find out new ways to maintain their lives.[24]

3.6 Political Life

The traditional Kom-Rem village had been ruled by a *Sawang* (Chief). The *Sawang* along with his male elders appointed by him, worked as a team in the administration of justice, enforcement of executive functions, maintenances of social practice and customary law etc.[25]

The village is the basic unit of communal living for the Kom tribe. Each village is ruled by a village chief or headman, known as *Sawang*. The village chief will have a council of elders to help him run the administrative system of the village smoothly. Normally, the position of the *Sawang* is hereditary although there are few instances where people elected their own *Sawang* for a certain period of time. The Manipur Village Authority Act of 1956 passed several Acts such as; permission of any village having not less than 20 houses to have village authority. Ultimately, the *Sawang* will be the chairman of the village authority.[26]

Sadar Hills Kom Union was formed in 1947 with the primary objective of uniting the five sub-tribes of Komrem such as Kom, Chiru, Aimol, Koireng and Purum. The Union was later changed into Kom Rem Union in the year 1957. Though, the Union does not have a separate political party, yet it renders a great help to many political parties as well as individual candidates.[27]

Young boys and girls remained under the control of their father at home but during *Lom* (Corporate Labour) in the village or outside, the *Lom-Upa* (Community Work Director) takes the responsibility. He looks after the conduct and behavior of the youths and if necessary disciplines them. Whenever they engage in a certain activity, they always followed the word of the eldest among themselves. For instance, whenever a group of people are engaged in fishing at the end of the day, they will club together all the caught fish and distribute. In distribution the biggest fish will goes to the eldest and the rest they will be distributed among themselves.[28]

If a youth is not able to control his/her social behavior and does not obey the *Lom-Upa* then the person is usually expelled from the society. The system of *Lom* or the corporate labour included cultivation, collecting fire wood, social service etc. *Lom* has been an organization for training young people for leadership in a village and also teaches a sense of duty and dignity of labour. Thus, *Lom-Upa* has been very important in the Kom-Rem society.

3.7 Cultural Life

Like some other tribes, the Kom-Rem people are a patriarchal tribe and the family lineage descends from the father's side. Ladies come and stay at their husband's house after the marriage. There are strict rules regarding marital relationships and those who cross the boundary line were excommunicated from their family lineage.[29] It was and still a traditional practice that if a man marries a woman from another clan, all their descendants should take the clan name of the father.

When a new baby is born, the woman folk of the village visit after a week with an egg each and some husk of rice to greet the new born baby, for a blessing of long life and prosperity. Whenever a person kills a tiger or any wild animal, the whole village would gather and sing their cultural and traditional songs over the death animal and later on they cook and eat together. The skull of the animal is then hung on the wall as this was a pride to the family and a great prestige in society. The Kom-Rem people believed that the evil spirits hated the blood of dogs, and thus the head of a dog is usually hung on the village gate to keep away the evil spirits from entering.

The Kom-Rem people were fond of drinking *Vai-zu* or liquor brew out of the rice mixed with husks and *Khazi-zu*. This *Khazi-zu* contains the same materials as that of the *Vai-zu*, but it is cooked on the fire. The pot is tightly shut with a small hole for putting a pipe which is connected to another pot. When it is heated, the filter distilled through evaporation is carried through the pipe to the other pot. The new collected distilled drink was called as *Khazi-zu*. Both these drinks were popular to both the gender.[30]

The Kom-Rem are fond of singing and dancing and perform various dances, like *Waikhong lam, Bechep lam* and *Sakhong lam* on annual festivals.[31] They use musical instruments like *Sirangdar* (a string instrument), *Rusem* (a wind instrument), *sum* (metal gong), *Khong* (drum) and *Dar* (small gong).[32]

3.8 Religious Life

Kom-Rem is a religious people and had their own religious system with a set of beliefs and practices. Separation between sacred and secular was foreign to them, so a study of their religiousness is a study of their whole life. Religion to them was what their gods did to them and their response to them.

The Kom-Rem people believed in one Supreme Being who is the creator of everything. He is called *Pathen* or *Chung Pathen*. He was believed to have stayed aloof after creation and involved in the lives of His creation only in cases of dire necessity.

There had also belief in the existence of many other powers including spirits who were considered more powerful than human beings. These spirits seemed to bring both good and evil to people. In order to ward off the evil influences of malevolent spirits, sacrifices were done to appease them.[33]

The *Thempu* (Priest) and *Thempi* (Priestess) had played an important role in these sacrifices. The *Thempu* stands as head of all in connection with religious rituals and sacrifices. The people accepted him as ordained by god and would go to him in the case of any disease and disasters.[34]

In course of time, the work of the Christian missionaries through schools had exposed a few students to Christianity. Such exposure paved the way for the growth of Christianity in the Kom-Rem traditional society, helping people to see the developments and changes of the outside world. This realization to a great extent built up their economic, social, political and religious life. Though majority of 95 per cent are Christians, about 5 per cent of the population still follows their old traditional faith and beliefs.

Endnotes

1. Snaitang, O. L. 2009. 'Christianity and Change among the Hill Tribes of Northeast India', in T. B. Subba, Joseph Puthenpurakal and Shaji Joseph Puykunnel (edited). *Christianity and Change in North-east India.* New Delhi: Concept Publishing Company. (P. 146).

2. *North-East India.* http://en.wikipedia.org/wiki/North-East_India. Accessed on 23 January, 2009.

3. *North-east Zone Cultural Centre.* http://www.nezccindia.org/introductionNE.asp. Accessed on 23 January, 2009.

4. Kamkhenthang, H. 1986. *In Search of Identity.* Imphal, Manipur: Kuki Chin Baptist Union. 17.

5. Singh, K. S. 1998. *India's Communities.* Vol. V. New York: Oxford University Press, 1785.

6. Kamkhenthang, H. *In Search of Identity* (P. 17).

7. Meitei are also call as the Manipuris, the largest ethnic group inhabiting in the state. They are group of people who Hindunised some 400 years ago. They dominantly live in the central valley and is surrounded by hill peoples.

8. Serto, Paoneikhup.1999. *A Study of the Religious Beliefs of the Kom-Rem Tribe before and after the Advent of Christianity.* (M.Div Thesis, Asian Institute of Theology). (P.1).

9. Kom, S. L. ed. 2002. *Kom Thrut Hla.* Imphal, Manipur: n. p. (P.13).

10. Kom, L. Benjamin.1990. *The Kom-Rem People.* Manipur: T. Yangpi Kom. (P.8).

11. Serto, Paoneikhup.1999. *A Study of the Religious Beliefs of the Kom-Rem Tribe.* 1.

12. Singh, K.B. 1976. *An Introduction to Tribal Language and Culture of Manipur: 7 tribes.* Manipur: Manipur State Kala Akademi. (P.63).

13. Kom, L. Benjamin. 1990. *The Kom-Rem People.* (P.7-8).

14. Kamkhenthang, H. 1986. *In Search of Identity.* (P.17).

15. Serto, Romeo.1972. *A Study of the Encounter of the Gospel with the Social, Economic, Political and Cultural Life of Kom-Rem People in India.* (B.D Thesis, Senate of Serampore) (P.7).

16. Kom, L Benjamin. 1990. *The Kom-Rem People.* (P.5).

17. Serto, Paoneikhup. 1999...... (P.3).

18. Kom, Serto Neilenkhup. 2009. *An Appraisal of Brahmabandhab Upadhyay's Understanding of Trinity from the North-eastern Perspective.* (M. Div. Thesis, COTR). (P.52).

19. Singh, K. B. 1976. .... (P.77).

20. Kom, T. Neilenshong. 2005. *A Study on the Cultural and Religious Beliefs of Keihrap-People Before and After the Advent of Christianity.* (M.Div. Thesis, Asia Theological Association). (P.6).

21. Serto, Romeo. 1972. .... (P.8).

22. Singh, K. B. 1976...... (P.66).

23. Kom, L Benjamin. 1990. .... (P.34).

24. *Ibid.:* (P.35-36).

25. Gangte. T. S. 1993. *The Kukis of Manipur .* New Delhi: Gyan Publishing House. (P.125).

26. Kom, L. Benjamin. 1990. .... (P. 49).

27. *Ibid.:* (P.48-49, 52).

28. Kom, T. Neilenshong. 2005. .... (P.7).

29. Serto, Romeo. 1972. .... (P.8).

30. Serto, Paoneikhup. 1999. .... (P.7).

31. Serto, Thangneireng. 2007. *Cultural heritage in Kom.* Imphal, Manipur: Tribal Research Institute. (P.62-64).

32. *Ibid.:* (P.55-61).

33. Vaiphei, Kim. 1995. *The Coming of Christianity in Manipur.* New Delhi: The Joint Women's Programme. (P. 4).

34. Serto, Romeo. 1972. .... (P.6).

References

Gangte. T. S. 1993. *The Kukis of Manipur .* New Delhi: Gyan Publishing House.

Kamkhenthang, H. 1986. *In Search of Identity.* Imphal, Manipur: Kuki Chin Baptist Union.

Kom, L. Benjamin.1990. *The Kom-Rem People.* Manipur: T. Yangpi Kom.

Kom, S. L. (ed). 2002. *Kom Thrut Hla*. Imphal, Manipur: n. p.

Kom, Serto Neilenkhup. 2009. *An Appraisal of Brahmabandhab Upadhyay's Understanding of Trinity from the North-eastern Perspective*. (M. Div. Thesis, COTR).

Kom, T. Neilenshong. 2005. *A Study on the Cultural and Religious Beliefs of Keihrap-People Before and After the Advent of Christianity*. (M.Div. Thesis, Asia Theological Association).

North-East India. http://en.wikipedia.org/wiki/North-East_India. Accessed on 23 January. 2009.

North-east Zone Cultural Centre. http://www.nezccindia.org/introductionNE.asp. Accessed on 23 January 2009.

Serto, Paoneikhup.1999. *A Study of the Religious Beliefs of the Kom-Rem Tribe before and after the Advent of Christianity*. (M.Div. Thesis, Asian Institute of Theology).

Serto, Romeo.1972. *A Study of the Encounter of the Gospel with the Social, Economic, Political and Cultural Life of Kom-Rem People in India*. (B.D Thesis, Senate of Serampore).

Serto, Thangneireng. 2007. *Cultural heritage in Kom*. Imphal, Manipur: Tribal Research Institute.

Singh, K.B. 1976. *An Introduction to Tribal Language and Culture of Manipur: 7 tribes*. Manipur: Manipur State Kala Akademi.

Singh, K. S. 1998. *India's Communities*. Vol. V. New York: Oxford University Press.

Snaitang, O. L. 2009. 'Christianity and Change among the Hill Tribes of North-east India', in T. B. Subba, Joseph Puthenpurakal and Shaji Joseph Puykunnel (edited). *Christianity and Change in North-east India*. New Delhi: Concept Publishing Company.

Vaiphei, Kim. 1995. *The Coming of Christianity in Manipur*. New Delhi: The Joint Women's Programme.

About the Author

Mr. Dongpu Kom is an Evangelist. He completed his Master in Divinity (M.Div), from New Theological College, Dehradun, India.

E-mail: doupu4christ@yahoo.co.in   Phone: 8377026759

Chapter 4

Traditional Prescriptive Marriage versus Cross-Cultural Marriage Systems among the Chothe of Manipur

☆ *Cheithou Charles Yuhlung*

ABSTRACT

This chapter highlights the significant of popular cross-cultural marriage among the Chothe with the neighbouring communities that poses threat to their traditional endogamous prescriptive marriage system. However, it indicates that despite the declined the system is still ideal, stable and adaptable. The declined of their traditional marriage system does not mean that the system is totally breaking down but rather a deviation. It indicates that the increased in cross-cultural marriage was develop due to their exposure, liberal worldviews and flexibility in their socio-cultural norms influenced by the thoughts of modern education, Christianity, westernisation and modernisation, etc. This liberal attitudes impacted by such factors allow a person to have more options in choosing a girl or a boy outside his/ her tribe. On the basis, a new trend of love and arranged marriage becomes popular a blend of tradition and modern marriage, common even among other North-eastern tribes of India.

Keywords*: Prescriptive, Cross-cultural, Marriage, Deviation, Chothe.*

4.1 Introduction

Chothe is an indigenous tribe of Manipur, predominantly inhabiting in Bishnupur and Chandel districts. They have their own distinctive language, customs, religion, marriage and kinship, economic and political institutions. Literature like Shakespeare (1912) have categorised Chothe as an Old-Kuki tribe of Manipur under culture and linguistic dimension. But politically they are now aligned with the Naga ethnic group. They also come under minority category because of their thin population. They practice patrilineal descent and patriarchal system. They practice "matrimonial cross-cousin alliance marriage system" as prescribed by their society that still persists even today. The Chothe though small had a significant place in anthropological world with debates for their distinctive endogamous prescriptive marriage system that began with Claude Levi-Strauss (1949) and Rodney Needham (1958-64) in reference to T.C Das (1945) work on ' *The Purum'* (Chothe).[1]

The Chothe, since time immemorial have been practicing their distinctive endogamous prescriptive marriage system which is now changing in the direction of popular exogamous form. This means there is sharp increased in the inter- or cross-cultural marriage among the Chothe with their neighbouring communities like Lamkang, Moyon, Tangkhul, Rongmei (Zeliangrong), Meitei, Hmar, Kom and others as compared before. The Chothe traditional prescriptive marriage system that was once considered as 'ideal, stable and adaptive', and as remarked by scholars like Needham (1958:97-98) against Livingstone (1959) hypothesis that someday this system would breakdown for such small society. The decline, however as expected is not in accordance to the actual Livingstone's hypothetical breakdown of the system but rather a 'deviation' or 'diversion' from the traditional endogamous prescriptive trend. The deviation is identified due to the influence of certain factors which compel them to adapt to a trend of love and arranged marriage. Under such circumstances there is increased in cross-cultural marriage among the Chothe as well as among other communities in the region. Thus, the paper indicates the emerging phenomenon of the increasing cross-cultural marriage based on love and arranged marriage.

4.2 Origin of Chothe

Mythically, the Chothe believe that the origin of the word *Shote - Zote - Chawte - Chote - Chothe* is derived from the word "Kachokte" or "Kachoite", literally it means 'the child that I held or stirred with'. Mythically, he is assumed the first Chothe. The second person is known as "Thanidam" (lit. the moon and the sun are alright) who became Kachoite's wife. They lived in a mythical cave called as *Huipi-thoranga,* meaning 'the cave from where the five brothers sprang' (Hiyang: 1985:15). Kachoite and Thanidam bore seven sons and two daughters who became the first progenitors or ancestors of Chothe. Etymologically, the name of these seven sons; thereafter becomes the seven clans of Chothe. They are (i) Khiyang/ Khiang/ Hiyang, (ii) Yuhlung or Zulung, (iii) Makan, (iv) Marim/ Mareem, (v) Thao, (vi) Parpa, and (vii) Rungkung/Rangshai. Each clan has its sub-clans. There are seven clans and seventeen sub-clans (Gupta 1985:74; see in Yuhlung 2005:49).

Geographical Location

The Chothe are concentrated in two districts of Manipur *i.e.* Bishnupur and Chandel. According to 2011 Census their total population is about 3,585 and literacy rate is 69.79 per cent , spread in twelve (12) recognised villages and five (5) new settlements. However, they may be divided into three regional groups based on their village agglomeration *viz.*: (1) Western, (2) Central-cluster and (3) Eastern groups (Basu 1985:38).

1. *The Western Chothe* comprises of two villages (i) Lamlanghupi and (ii) Lamlanglon (new settlement). Lamlanghupi is considered the parent village of all the Chothe tribe, since the rest of Chothe is said to have scattered from the western hill range. The Census of India still recognizes Lamlanghupi as "Chothe-Munpi" recorded under Churachandpur district of Henglep Constituency.[2] These villages are about 2 kms from Bishnupur town and 27 kms from Imphal.

2. *The Central-cluster Chothe* is located on the extreme western Hirok hill range of Chandel district. It comprises of nine prominent villages and five new settlements. Depending on the nearness to the market from their villages they do their shopping in the towns of Kakching, Pallel and Chandel Bazaars.

3. *The Eastern Chothe* comprises only one village called Khongkhang. It is located on the Indo-Myanmar National Highway road (NH-39) after crossing the Tengnoupal peak of Chandel. The village is about 76 km from Imphal, 32 km from Pallel and about 26 km away from Moreh town.

4.3 Language

Chothe language is their mother tongue. Although there is a dialect variation among regional groups but it is intelligible to all. Most Chothe can fluently speak Manipuri or Meitei language being the Langue-franca of the State. Since, Chothe belong to Mongoloid racial stock, linguistically their language belongs to the family of Tibeto-Burman under the sub-group of Chin-Kuki-Mizo speakers.[3] They have no script of their own. They used either the Roman script or Bengali-Manipuri script in writing. Their history is based on oral narratives that have been passing down from one generation to another over the ages. Their language and culture is very close with tribes like Kharam, Koireng, Kom, Chiru, Aimol, Lamkang, Tarao, Puimei, Liangmei (a sub-group of Zeliangrong), besides the New-Kuki or Thadoun and Mizo speaking groups. The next groups that come closer linguistically are Anal, Moyon, Monshang and Maring.

4.4 Neighbouring Communities of Chothe

The Western Chothe is immediately surrounded by diverse ethnic groups like Rongmei (Zeliangrong), Kom, Chiru, Aimol and Kuki (Thadou speaking groups) in the west and southern part while the Meitei are found on the north and east.

The Chothe Central-cluster is surrounded by various ethnic tribes like Anal, Moyon, Monsang, Lamkang, Tarao, Maring on the east and south while the Meitei on the North and west.

The Eastern Chothe neighbouring tribes are the Maring, Lamkang, Kom and Kuki (Thadou speaking groups).

4.5 Kinship Structure

The Chothe adopts patrilineal descent, patriarchy and patrilocal systems. They also practiced matrilateral cross-cousin (MCC) marriage, characterised by the principle of 'endogamous prescriptive marriage' system. In olden days, the Chothe family was characterised by joint or extended family but nowadays under the influence of modernisation and westernisation many have adopted to nuclear family. The eldest male child is the successor as the head of the family after his father's death (not the youngest). So, he is referred as *Pipa/Shapa* the 'male head of the family/ lineage/ clan'. But in term of inheritance of family's property the youngest son is the ultimogeniture. He inherits his parent or grandparent's property. Although in certain cases, some immoveable family's properties like land are shared equally among brothers except the parent's share. The girls are known as *Sarrnu/ Sheinu* (spear). In tradition, girls are not entitled to inherit any family's property. But these days those girls belonging to rich and wealthy family received certain amount of family's property in the form of gifts.

Prescriptive versus Cross-Cultural Matrimonial Alliance System

The tribe became popular in social anthropology during 1950-1970, after Claude Levi-Strauss' put forward in his theoretical paradigm of, '*Les Structures Elementares de la Parente*' (1949) citing Das work '*The Purums*' (1945). After that various eminent scholars like George C. Homans and David M. Schneider (1955), Rodney Needham (1959-1960's), Frank B. Livingstone (1959), Floyd G. Lounsbury (1962), Charles Ackerman (1964), William H. Geoghegan and Kay Paul (1964), George L. Cowgill (1964) and William Wilder (1964) and various others debated and argued on the Chothe prescriptive marriage system, against on the backdrop of 'alliance theory' and 'matrilineal connubium'. Scholars like Robin Fox and others have referred Chothe marriage under 'asymmetrical and symmetrical' of complex marriage system in their studies (1967: 210).

Parallel-cousin marriage or marriage with father's brother's daughter or father's sister's daughter (FZD) is forbidden among Chothe. It is also considered taboo since viewed as cognate group or the bloodline being too near. On one such ground, they adopted a specific endogamous prescriptive matrimonial alliance system in which matrilateral cross-cousin (MCC) or marriage with mother's brother's daughter (MBD) is the most preferred form. This preferential gave rise to prescriptive marriage system probably after prolong practice in society. Thereby as accepted by the society, a system of prescriptive matrimonial alliance system had developed, accordingly the rule is that a Chothe boy should marry a girl only from the two or three prescribed clans and its sub-clans, out of the existing seven clans and fourteenth sub-clans (also see Yuhlung 2007).[4] No girls should be directly exchange.

According to this endogamous prescriptive marriage rule the Chothe exchange their girls in circle. They forbid direct exchange of girls between lineage or sub-clan or clan viewing as cognate. Taking girls from their un-prescribed lineage or clan is strictly considered a breach and against their marriage rule. In short, their marriage rule is that a boy cannot take a girl from the clan or sub-clan he gave his sister. In this way, the girls flow in one direction. Therefore, Needham (1958) said that the Chothe marriage functions in connubium style where girls are exchange indirectly among specific lineages or clans and operates in cyclical or rotation manner. In the Chothe society no specific clan or sub-clan is superior or inferior. The marriage system operates in a cyclical manner, where a clan accepts a girl from another clan or sub-clan, and gives its girls to yet another clan or sub-clan.[5]

The indirect exchange of girls or circle system is possible among the Chothe because of the triadic structure as: (i) wife-giving, (ii) wife-taking and (iii) unrelated or distant relative or neutral group. Das and Needham also explained as: (i) a boy's or rather his father's sib, (ii) his mother's group of sibs, or rather the group from which his wife is recruited, and (iii) his sister's husband's group of sibs, - this is sometimes his mother's mother's or his mother's mother's mother's sib.[6] Therefore, this allows a person to choose a girl from the other two unrelated groups or prescribed clans.

On the above basis, the Chothe boys choose and marry girls either i) from his MBD, or ii) from his mother's clan, or iii) from other prescribed clans or if not from outside his community. Table 1 shows how the Chothe still conforms to their age-old traditional prescriptive matrilateral cross-cousin marriage besides indicating the increased cross-cultural marriage with other communities (also see in Yuhlung 2007:52-53). Out of twelve villages and five new-settlements (12+5 = 18), with a total population of 2689 and a household of 531 there were a total of 591 marriages, out of which 446 were girls from within the Chothe tribe, while 145 were girls belonging to different tribes or communities. Another significant aspect of this data (Table 4.1) is that the Chothe still adheres to their prescriptive marriage rules, where out of 446 marriages, 37 girls actually were girls of Mother's brother's daughter (MBD), and 103 were from the prescribed mother's clan, which shows that their traditional marriage system is 'ideal, stable and adaptive' till today, despite the deviation with the increased in cross-cultural marriage from their traditional.

In the earlier stage, the cross-cultural marriage was initially initiated by only few people who faced resistance from family and society. But today it has become a very popular fashion among the Chothe and also even among other North-eastern communities, so also around the world. Especially in the west, there is an increase inter or cross-cultural marriage among diverse communities of the world where most of the present generation are the descendent of mixed races or cultures of; Black and White, White and Brown, Black and Brown, etc. There are also increased inter-religious marriage taking place around the globe like between Hindu and Christian, Jew and Christian, Muslim and Christian, Buddhist and Christian, and so on. Thus, it means girls are increasingly exchanged more freely among diverse communities irrespective of caste, class, creed, religion, race, culture or tribes unlike in the past. People are now seen to have changed their mindset liberalising from close to open because of globalisation and advancement in science and technology.

According to Y. Lungle (37/M) and Pr. Hiramani (40/M) of Tampakhu village the preference for cross-cultural marriage is that "In general, the offspring of most cross-cultural parent are better looking, beautiful or handsome, smart and intelligent

Table 4.1: Chothe Matrimonial Alliance System (Based on field data collected in 2004)[7]

Name of the Village	Total No. of Household	Total No. of Population	No. of Clans and Initial of Clans in a Village	Total No. of Married Couples	No. of Marriage with MBD	No. of Marriage with Mother's Clan	No. of Marriage with same Tribe	No. of Marriage with others Tribes
Ajouhu (Purum Khullen)	38	164	(7) K, M, Mk, Y, Th, Pr, (Rs)	35	7	8	28	7
Bethel Happy land	04	13	(4) K, M, Th, Pr.	04	Nil	Nil	4	Nil
Chandolpokpi	19	91	(4) K, Mk, Y, Th.	22	2	4	16	6
Chothe Khunou	14	67	(5) K, M, Mk, Y, Pr.	15	3	8	12	3
Chumbang (Purum)	72	336	(6) K, M, Mk, Y, Th, Pr.	79	2	10	61	18
Khongkhang	49	296	(6) K, M, Mk, Y, Th, Pr.	62	1	7	56	6
Laininghu	35	204	(5) M, Mk, Pr, Th, Y	43	3	10	38	5
Lamlanghupi	53	291	(7) K, M, Mk, Y, Th, (P), (R).	50/58	Nil	Nil	24	34
Lamianglon	13	42	(3) M, Th, Y.	13	NIL	Nil	2	11
Leirungtabi	05	27	(4) K, M, Mk, (Rs).	05	Nil	Nil	2	3
Lunghu	06	39	(4) M, Mk, Y, Th.	09	Nil	4	7	2
Lungleh	10	50	(3) M, Mk, Pr.	10	Nil	Nil	5	5
New-Wangparal	25	138	(4) M, Mk, Y, Pr.	23	5	2	17	6
Old-Wangparal	21	111	(4) K, M, MK, Th.	24	5	12	21	3
Phantu (Chandrapoto)	54	260	(6) K, M, Mk, Y, Th, Pr.	61	7	13	49	12
Salemthar	10	42	(3) M, Mk, Pr.	10	Nil	Nil	6	4
Tampakhu (Purum)	72	332	(7) K, M, Mk, Y, Th, (P), Pr.	83	1	22	69	14
Ziontlang	31	186	(5) M, Mk, Y, Th, (Rs).	35	1	3	29	6
Grand Total	531	2689		591	37	103	446	145

Abbreviations of clan and sub-clan names: K- Khiyang/ Hiyang, M-Marim, Mk-Makan, Y- Yuhlung, Th- Thao, Pr- Parpa, R- Runkung, P-Pilling, Rs- Rangshai.

Source: Yuhlung (2007:52-53).

compared to the endogamous marriage partners, in this way we are in favour of cross-cultural marriage, besides expanding the relationship beyond our tribe". On the other hand, Mrs. Mk. Pishak of Pallel and other likeminded Chothe women opines that, "marrying girls from the endogamous prescriptive groups are far better in the long run. They are mostly successful in their marriage and compatibility compared to the exogamous or cross-cultural marriage that often resulted in divorce, broken marriages and conflicts". Both the views seem to hold true and so it is left to individual choice in selecting which types of girls one marries.

However, a contradiction that discourages this popular cross-cultural marriage is observed among few exceptional tribes like the Mizo, Bhutanese and Tibetans. It is learned that these communities strongly discourages it by propagandising politically and at various religious gatherings. Kinley Dorjee from Paro, Bhutan said that "the Government of Bhutan had carried out an ethnic cleansing by 2005 because there were too many illegal immigrants like Nepalese, especially coming from the west. After expulsion their population was considerably reduced. It is also to do with our belief to maintain the pure cognatic blood group from being polluted with other outsiders".

Though, science claimed that marriage with cognate or consanguine resulted in certain physical deformation and other detrimental effects. However, despite the heresy, they encouraged and recommend for endogamous marriage system whereby discouraging the exogamous cross-cultural marriage for various reasons like to stop the influx of foreigners or as ethnic cleansing or as part of the endogamous culture to retain its pure blood. But with the passage of time and the arrival of post-modernity inter/ cross-cultural marriage becomes a popular trend accepted worldwide. Today, even the most rigid society in the world like Chothe have adapted to such inter-cultural marriage trend without much opposition.

In Manipur, so far by the Census of India 2011 there are 33 recognised scheduled tribes and various sub-tribes, besides communities like Meitei, Meitei-Pangal (Muslim), Scheduled castes and many other sub-groups. Table 4.2 shows out of 591 total marriages 154 were exogamous marriages or marriage with girls from other communities which means the Chothe have selected girls atleast from twenty five (25) tribes or communities, who conform to the traditional and modern (love) and arrange marriage. However, keeping in mind the variation of number of exogamous marriage in connection to Table 4.1, it was later reported that nine (9) girls had divorced between the time of data collection and my earlier publication on 'Matrilateral cross-cousin marriage among the Chothe of Manipur'(2007). So, the actual number of exogamous marriage during that time may be considered as 145, but this number is expected to have risen by now with increased population, exposure, liberalisation, globalisation and flexibility in the social systems.

The significant of this data in Table 4.2 is that, for obvious reasons like nearness, similarity with one's culture, like mindedness etc. the village with the highest 36 exogamous marriage is Lamlanghupi followed by Chumbang with 20, Tampakhu with 14, Phantu with 12, Chandolpokpi and Lamlanglon with 11 each and Ajouhu with 7. The rest of the villages had below six (6) exogamous marriages.

Table 7.2: Women of different Tribes Married to Chothe Men (Actual)

Sl.No.	Women of different Tribes	1	2	3	4	5	6	7	8	9	10	11	12	13	14	15	16	17	18	19
																*Name of the Chothe village****				
1.	Aimol	–	–	–	–	–	01	–	–	–	–	–	–	–	–	–	–	–	–	01
2.	Anal	–	–	–	–	01	–	–	–	–	–	–	–	–	02	–	1	–	–	04
3.	Gangte	–	–	–	–	–	–	–	01	–	–	–	–	–	–	–	–	–	–	01
4.	Hmar	02	–	–	–	–	–	2	03	–	–	–	–	–	–	–	–	–	1	08
5.	Khasi	–	–	–	–	–	–	–	01	–	–	–	–	–	–	–	–	–	–	01
6.	Kom	–	–	–	–	03	–	–	04	–	–	–	–	03	–	–	–	–	1	11
7.	Lamkang	–	–	03	02	04	–	–	–	–	02	–	1	–	–	03	–	02	–	18
8.	Lushei (Mizo)	01	–	–	–	–	01	–	01	–	–	–	–	–	–	–	–	–	–	02
9.	Mao	–	–	–	–	–	–	–	–	01	–	–	–	–	–	–	–	–	–	01
10.	Maram	–	–	–	–	01	–	–	–	–	–	–	–	–	–	–	–	–	–	01
11.	Maring	–	–	–	–	–	01	–	–	–	–	–	–	01	–	–	–	–	1	03
12.	Meitei	02	–	01	–	02	–	–	01	02	–	1	1	01	01	01	1	04	1	17
13.	Monshang	01	–	–	–	–	–	–	–	–	–	–	1	–	–	–	–	–	–	02
14.	Moyon	–	–	01	–	–	03	–	01	–	–	1	–	–	–	01	–	–	–	09
15.	Nepali	–	–	–	–	02	–	1	–	–	01	–	–	–	–	–	–	–	–	04
16.	Paite	–	–	–	–	–	–	–	–	01	–	–	–	–	–	–	–	–	1	02
17.	Poumei	–	–	–	–	–	–	–	01	–	–	–	–	–	–	–	–	–	–	01
18.	Rongmei	–	–	01	–	–	–	–	19	06	–	–	–	–	–	01	1	03	–	32
19.	Simte (Zou)	–	–	–	01	–	–	–	01	–	–	–	–	–	–	–	–	–	–	01

Contd...

Table 7.2–Contd...

Sl.No.	Women of different Tribes	Name of the Chothe village**																		
		1	2	3	4	5	6	7	8	9	10	11	12	13	14	15	16	17	18	19
20.	Tangkhul	–	–	04	–	03	–	–	02	01	–	–	–	–	–	05	1	–	1	17
21.	Tarao	–	–	–	–	02	–	–	–	–	–	–	1	–	–	–	–	–	–	03
22.	Thadou (Kuki)	–	–	–	–	02	–	1	–	–	–	–	–	01	–	–	–	05	–	09
23.	Tripuri	–	–	01	–	–	–	–	–	–	–	–	–	–	–	01	–	–	–	02
24.	Vaiphei	01	–	–	–	–	–	1	–	–	–	–	1	–	–	–	–	–	–	03
25.	Others/ Loi (caste)	–	–	–	–	–	–	–	01	–	–	–	–	–	–	–	–	–	–	01
	Grand Total	07	–	11	03	20	06	5	36	11	03	2	5	06	03	12	4	14	6	154

**: 1: Ajouhu (Purum khullen); 2: Bethel Happy Land*; 3: Chandolpokpi; 4: Chothe Khunou; 5: Chumbang (Purum); 6: Khongkhang; 7: Laininghu; 8: Lamlanghupi; 9: Lamlanglon*; 10: Leirungtabi; 11: Lunghu*; 12: Lunghu*; 13: New-Wangparal; 14: Old-Wangparal; 15: Phantu (Chandrapoto); 16: Salemthar*; 17: Tampakhu; 18: Ziontlang; 19: Total no. of each tribe.

The *(asterisk) marked six villages are new settlements. Out of a total of 154 marriages with outside girls, 09 of them were divorced during the time of data collection.

Another significant aspect is that the Chothe received a maximum of 32 girls from the Rongmei (Zeliangrong) tribe, followed by Lamkang with 18, Tangkhul and Meitei with 17 each, Kom with 11, Moyon and Thadou speakers (New Kuki) with 9 each, and Hmar with 8. The rest of the communities are below seven (7) girls. This reflects that, as the first option, the Chothe select girls especially within their neighbouring communities. Secondly, took girls mostly from the acquainted communities. Despite the increased of popular cross-cultural marriage many Chothe still upholds their traditional endogamous prescriptive marriage system till today without directly breaking their prescriptive marriage rule. The increased exogamous marriage does not mean that Chothe prescriptive marriage system has been completely broken down but rather shows the decline and deviation from their traditional endogamous marriage system because of the liberal globalisation phenomena, thereby leaving more ample rooms in choosing a girl from varied groups.

4.6 Marriage Forms

The Chothe broadly practices three types of traditional marriage system *viz.*: (i) *Nu-ngak loh* (Arranged marriage), (ii) *Mou-sem* (Love marriage) and, (iii) *Tlang-chom neilah* (Elopement/ Force marriage).

This *Nu-ngak loh* marriage literally means 'wooing a girl by visiting/ staying' at her parent's house. This is a kind of prenuptial engagement or preparatory or arrangement marriage. It is widely practice in early days by common people where the boy agreed to serve three years marriage labour service by visiting or staying in the girl's house in the form of paying the bride-price. Often, the girl here is the mother's brother's daughter (MBD) which is the most preferred and prescribed by the society. In the past, this is the most common and preferred form of marriage among the Chothe for its idealness, stability and adaptability.

For Chothe, marriage is an important institution in building the kinship alliance relationship. In case, there are three-four male child in a family at least one male child must marry his actual MBD to continue their alliance relationship. In case, a person's mother's brother bore no female child then he may choose it from his mother's brother's lineage or mother's clan. Traditionally, perhaps if none of the male child does not marry his MBD, it is considered a kind of disgraced and humiliation to his mother's family or lineage or clan for terminating the marriage alliance. Y. Hongpa (my informant) said that in the past, under such circumstances his mother's brother's family or lineage imposes a fine; a pig on two grounds. One reason is for disgraced to his mother's brother's lineage and the second for terminating the marital and kinship alliance relationship between the two families or clans.

The second marriage is known as '*Mousem*' or '*Ruihong*' which means 'preparing or bringing a bride'. It is a kind of both love and arranged marriage. It is a marriage mutually solemnised when the boy finds a girl of his choice or his soul mate from within the prescribed clan or outside the community. Often, the boy is from royal family or higher status. In other words, it is a royal marriage (of kings and princes). The boy proposes the girl and seeks the consent of her parent for a kind of direct marriage by paying the required bride price, without serving the traditional three years marriage labour service. When both the parties mutually agreed the terms and conditions of their love marriage an auspicious day is fixed for the wedding. On such auspicious wedding day the boy pay the bride price by offering; a pig, a basket of boiled meat (cow or buffalo), a pot of liquor *(zu),* besides other essential items like a spear, a knife, and three traditional shawls and rupees five hundred in coins to the girl's family. This tradition of giving of bride price is known as '*maansipa*'. Here, only the prescribed and close paternal kins of the bride are allowed to eat the meat and drink the wine brought by the boy's family. Other members are restricted to partake but normally another separate pig or cow may be killed as part of the marriage feast where all expenditures are borne by the boy. This *mousem* marriage is also known as *maan-loh,* meaning 'demanding the bride price'.

The third type of Chothe marriage is known as '*Tlaan-chom neilah*' meaning 'marriage by elopement'. It may be further sub-divided into two categories: (i) consent of two lovers to elope against the wishes of their parents, and (ii) forced elopement against the wish of the girl (force marriage) either by kidnapping or seducing by magic. The first category usually occurs when a boy falls in love with a girl outside his prescribed clan or village or community and the girl reciprocated the invitation. Such marriage is common among the youth when both parents refused to consent the marriage on certain personal reasons. The second category occurs among few youth or powerful or influential persons who kidnapped or seduced the girl they desired using magical love charm into force marriage. Pr. Roushi (formerly a village priest of Ajouhu) said that "the tribal in general seldom used love charms or witchcraft to win the heart of a girl, but this does not mean that they never use it". However according to him, "such love charms are commonly use by the Meitei now, which is why many wear amulets (*Chandra*) to protect themselves from such witchcraft acts".

According to the Chothe customary law in relation to the third type of marriage, if a Chothe girl elopes with a boy outside her prescribed clan or village or tribe, the boy has to pay a heavy fine for the breach committed and disrespecting their endogamous marriage rules. This cultural aspect of imposing fine for the violation of marriage rule is called as *Phung tlaanglam,* meaning 'fine for marriage outside the prescribed clan' or village or tribe. The penalty fee is a pig (five *wei* or *wai* meaning the pig's neck must be not less than 2.5 ft in diameter) and a bottle of wine that should be given to the Village Council for the breach committed. The girl's family members are prohibited from eating the meat and drinks brought by the boy as fine.

Interestingly, this elopement is the most common marriage form among the Meitei of Manipur, known as '*Nupi Chainba*' (girl being eloped). They also perform a kind of sacramental marriage known as '*Luhongba*'. Some Meitei elders said that 'elopement is cheaper and better form of marriage if one is not wealthy and rich'. In certain cases, when family complications are settled amicably, soon sacramental/ holy marriage may still be performed. The first cross-cultural marriage may be traced to the mother of King Pakhangba the first historical king of Manipur. It is mentioned that Pakhangba's mother was a Chothe girl called Sorha, and known by Meitei as Leinung Yabi Chanu (see Shaw 1929, Yuhlung 2011). It is believed that when an early group of immigrant from the west entered Manipur and occupied a nearby Chothe village a youth fell in love with one of the chief's daughter. Thereafter, it is believed that the indigenous groups and outsiders began to have cross-cultural marriage but elopement became only the means to get a tribal girl by others because of the rigid cultures during those early days. The outsiders took the advantage of the tribal condition because once a girl is being eloped without the approval of her parent she will be excommunicated for the breach, after the boy was penalised with a simple fine only. Here the objective of the boy is fulfilled in getting the girl, though marriage is invalid.

4.7 Marriage Solemnity

To consider Chothe marriage legal and valid one has to undergo different ceremonial stages. They have no specific written record about their marriage rules but it is deeply embedded in their oral tradition and customs. In case, if a boy marries a girl from outside his community the boy should conform to the girl's custom and tradition. On the other, whether it is arranged marriage or love marriage or elopement, in case a Chothe girl marries outside her village or community, the boy has to fulfill all the required customary rules accordingly like *Phung-tlanglam, Maanshipa, Loukhat-pa (Ruihong-pa).* Until and unless these formalities are fulfilled the marriage is considered illegal and invalid.

However, assuming that the boy has completed performing the first two important marriage ceremonies; *Phung-tlanglam* and *Maanshipa,* then only, he can performed the third and final *Loukhat-pa* the 'blessing/ sent-off ceremony'. Thus, the marriage becomes legal and valid only after receiving this girl's parents' blessing and were sent-off with some gifts (moveable property). Henceforth, the boy came to be recognised as *Amak-pa* or *Maksa* which means 'son-in-law' or 'the alliance partner'. The term is used and address exclusively by the girl's family/ lineage/ clan members only to the person who marry their girl to signify the specific marriage alliance connection between two families as - wife giver and wife taker.

In certain cases, if the information of elopement or forced marriage of a girl is not immediately notified to her parent within two-three days, the girl's family (if higher status) with the help of the villagers has the full legal right to avenge or wage war against the boy's family or village or community on the pretext of stealing their girl, or for daring and bringing disgraced to the girl's family/ clan or even to the village. This tension of war or conquest or feud is only calm down by immediately paying the fine *Phung-tlaanglam.* Only when this fine is accepted negotiation can take place between the two parties. In certain cases, if this fine is denied by her father, the boy may be forced to return the girl immediately.

In certain extreme cases, Y. Damshu said that 'even if the penalty fee was paid by the boy, sometimes the girl's father out of utmost anger may refuse to reconcile and solemnise the other official marriage ceremonies for the breach or going against his wish. In such cases, the couple live an invalid married life. All the legal rights concerning succession and inheritance may also be denied as they fail to comply and establish the alliance partnership, and received the blessings. Some girls were even excommunicated from her family and kin groups. Such was the rigidity of the Chothe culture and marriage system in the past.

4.8 Bride Price (*Maanshipa*)

In Chothe the practice of giving bride price by a boy to the girl's parent is called as "Maanshipa" meaning 'giving the bride price/ value'. The concept of taking or giving of bride price is different from the Indian dowry system. For Chothe the most important symbolic meaning of taking bride price is to acknowledge the trust, commitment and loyalty of the boy to the girl's family, to the extent of giving his life as alliance partner (*amakpa*), with the assurance to love, care, protect and support the girl as his wife till death. Therefore, the boy on that specific day swears by the spear, knife or gong in the name of his forefathers. Secondly, it signifies the high value of a girl, who will procreate his descendants to be remembered. It also means the most precious non-material asset of the father that he can give to someone worthy enough who can continue the life with fame, popularity and richness (for Chothe girls are not viewed as material object at all). Finally, another significant point is the establishment of a new alliance partnership or the continuation of the alliance relationship by marriage (in case one marries his MBD). So, in the context of Chothe the bride-price is a symbolic object to signify that the boy have assured to be loyal and even ready to sacrifice his life to the girl's family as an alliance partner while good care is responsible for the girl. For Chothe, loyalty is held higher than money, therefore it is not the money or gift that is important but it is the loyalty and trustworthiness between one another what matters most in friendship. On the positive note, perhaps suddenly a conflict arises in a village among different clans for certain reasons but if a family has three or four girls and are all married to different clans within the village, sometimes the conflicts are solved amicably in certain cases. Since, all are indirectly related by marriage it helps in resolving the conflict in society.

4.9 Factors Responsible for Deviation

There are certain factors identified responsible for deviation from the endogamous prescriptive marriage system. In the past, the arranged marriage (*Nungak lohpa*) or the endogamous prescriptive marriage system was the most common whereas love marriage comes in second place and at the third is the elopement (force marriage). But today, with the gradual improvement in the socio-cultural relationship and economy of the people through various agents and influence of changes the understanding and cooperation level among diverse ethnic communities in the region have tremendously ameliorated from their rigid fundamentalist outlook. Some prominent factors of influence of change are like; modern education, impact of Christianity, westernisation and modernisation, various government policies and programs, numerous organised local exchange programs like youth development, cultural exchange programs, annual local/ regional games and sports' meet, social ceremonial functions (weddings, funeral, feasts), inter-cultural church programs and fellowships, modern market place, liberalisation and globalisation, etc. enhances the free interaction and awareness amongst diverse communities. All these elements allow and provide a room for a person to change his or her mind set from traditional fundamental outlook to liberalism.

There is no doubt to deny that even prior to India's Independence (1947) there were inter- or cross-cultural marriage but the degree is seen to be less among the

Chothe as well as among other communities as compared today. However, the number of cross-cultural marriage has increased since India's Independence and the momentum in North-eastern region began from 1970's onwards with the introduction of modern education and Christianity. The pioneers who initiated inter-marriage were mostly educated people coming from different backgrounds. Majority of them had received *modern education* from different education institutions either run by the Government or private within and outside the state. These people because of their exposure, new experiences, and broader worldviews, and with jobs compel them for a different type of social relationship outside their traditional domain. It is said that initially, they were faced with opposition from the traditionalist for going against their prescriptive marriage rule but gradually people began to accept due to the growing number of inter-marriages even among other North-eastern tribes or communities. It is also said that even some youth who are the children or closely related to the traditionalists or members of village authorities continued to take or give their girls from outside their own group, despite resistance. On such conditions, the Chothe elders gradually became lenient and flexible in their societal norms and marriage rule in order to reduced the tension and conflict and conform to the existing situation.

Another important factor is the coming of *Christianity* in the North-east India. Christianity has had played a very vital role in changing most of the tribal societies of the region from narrow, pessimistic and rigid outlook to broader and optimistic worldviews. Christianity set its foot in Manipur on 6th Feb. 1894 first in Imphal but for political and cultural reason it moved to Ukhrul and sowed the seed of Christianity among Tangkhul Naga in 1895 by William Pettigrew an American Baptist Missionary (Sangma 1987: 13). Christianity has brought western education and western culture that has greatly shaped and change the mindset of the North-east people. Often the Christian (denominational churches) organised frequent combined worship services or fellowship programs, sometimes inter-village or inter-communities. Such programs and activities conducted provide space for free interaction and flexibility between villages or among neighbouring tribes, thereby reducing the rigidity in their societal norms, and giving more opportunity for a boy to select a girl outside his village/communities.

When North-east India became part of India after her Independence, along came the influence of modernisation and westernisation with the introduction of various developmental jobs, policies, programs and schemes to improve the living condition of the people. Such developmental policies and schemes also facilitate change in social institutions since all have to abide by the Constitutional law. Such avenues created more exposure and awareness with the surrounding tribes and community through frequent interactions and common activities.

The development of modern market in towns and sub-towns is another important factor for the increase of cross-cultural marriage since it served an ideal place for various groups of people to come together and interact freely. Such places provide an opportunity for a youth to find, select and also date a girl belonging to different group (while shopping or on the pretext of shopping) and finally may marry the girl of his choice.

The annual regional or inter-village games and sports meets organised also provides the youth to acquaint with new friends and exchange their views and ideas. Thereby, it increases the option to choose a girl of his choice from various sections of the group.

The social ceremonial functions like festivals, merit feasts, weddings, funerals and other related social functions also provide the best opportunity for the youth to meet new people and interact. Such occasions provide better opportunities for distant relatives or kins or friends of the host who came to attend from far and near places. The social gathering programs conducted on such occasion like feast, especially in the evening gave an extra opportunity to mingle and interact or strengthen their new bond of relationship. Besides many other, these are some factors identified for the increased in cross-cultural marriage among the Chothe, and also among communities like Meitei, Tangkhul, Anal, Moyon, Monshang, Lamkang, Meitei, Rongmei, Kom and others as observed.

In connection to the factors of influence and opportunity, the traditional arranged marriage that was very common in earlier days began to decline and was replace by love marriage (marriage by one's choice, initially it was often in the form of elopement). But because of the growing phenomenon of love marriage the structure and customary law of society was forced to change from rigid mind-set to flexible outlook. As a result, a serious debate seems to have risen before between the protagonist of traditionalist and modernist within the community. But, since none of the party could champion the debate; an agreement seems to have reached that it should be a combination of both tradition and modern (western style). Since, majority of the Chothe are now Christians they have conformed to this new marriage rule of "Love-Arranged marriage/ Arrange-Love marriage" a dual marriage ceremony. In this way, any love marriage among Chothe is now initiated by the traditional customary marriage norms like marriage proposal, negotiation to the girl's parent, penalty fees, and paying the bride price, etc. Only after completion of these traditional formalities a person may go for the western Holy marriage ceremony. Thus, a new marriage system has been developed which is a blend of traditional and western form *i.e.*, a combination of traditional (*Mousem* or *Ruihong* a kind of mutual love marriage) and Holy marriage.

Thus, any person marrying a Chothe girl should first complete the traditional marriage norms, and then only, one can proceed for the modern Holy marriage or any other forms of marriage ceremonies. But in case both the partners are non-Christians, only the traditional marriage rule needs to be fulfilled. The dual or cross-cultural marriage ceremony is gaining popularity since both the parties wants to solemnised the wedding according to their tradition and customs defined by religion, caste, tribe. Therefore, we have seen a number of young couples from North-east with different cultural backgrounds like a Christian girl marrying a Hindu boy solemnised their wedding twice according to their religion and culture. It is believed that this popular cross-cultural or exogamous marriage which is increasing among the Chothe and other North-eastern tribes is expected to continue for some more time before it falls back to their traditional endogamous system for certain unknown reasons.

Endnotes

1. T.C. Das 1945, '*The Purums: An old-Kuki tribe of Manipur'*, is a classical book that exclusively studies only the Central-cluster Chothe by disassociating the other two groups of Western and Eastern Chothe. The book had been forced from re-publication for the misnomer title since it has created confusion with the identity worldwide. This confusion of identity had been rectified by a re-visit team among the same people headed by Biman, K. Das Gupta (1985).

2. *Chothe-Munpi*: It refers to an old abandon village of dominant Chothe village of 17th century near the Leimatak river valley or Leimatak Hydro Electric Project.

3. See in Greirson, George Abraham. 1967.

4. See in Das Gupta, B.K (1985: 74)

5. See in Yuhlung (2007: 61).

6. See in Das, T.C (1945: 123, 125); Needham (1958: 80, 81).

7. Both the statistical data of table 1 and table 2 are personal census survey data collected during September-December 2004 fieldwork, while pursuing my PhD, which titled as "*Indigenous religion of the Chothe of Manipur: A sociological study*" submitted to the Department of Sociology of North-eastern Hill University (NEHU), Shillong in 2010.

References

Ackerman, Charles. 1964. 'Structure and statistic: The Purum case', in *American Anthropologist*. Vol. 66. 53-65.

Ansari, S. A. 1991. *Manipur: Tribal demography and socio-economic development*. Delhi: Anil Mittal.

Basu, Arabinda. 1985. 'Contemporary demographic pattern of the Purum (Chothe) with special reference to fertility and mortality', in Das Gupta K. Biman (ed.) 1985. *Proceeding on the symposium on: Purum (Chothe) revisited*. Calcutta: Anthropological Survey of India. Census of India, 2011.

Cowgill, L. George. 1964. 'Brief communications: The Purum case: More structure and Statistic, a critique of Charles Ackerman analysis of the Purum', in *American Anthropologist*. Vol. 66. 1351-1365.

Das Gupta, Biman, K. 1985. *Proceeding on the symposium on Purum (Chote) revisited*. Calcutta: Anthropological survey of India.

Das, Tarak Chandra. 1945. *The Purums: An old-Kuki tribe of Manipur*. Calcutta: Calcutta University Press.

Fox, Robin. 1967. Kinship and Marriage: An Anthropological perspective. UK: Harmondsworth, Penguin Books.

Greirson, George Abraham (Ed.). 1967. *Linguistic Survey of India*. Vol.III. Tibetan-Burman Family, Part-III (Specimens of the Kuki-Chin and Burma groups). Calcutta: Superintended of Government Printing), [reprinted, Delhi: Motilal Banarsidass].

Hiyang, Thambaljao.1985. 'The original custom and culture of Chote tribe', in Das Gupta Biman, K. 1985. *Proceeding on the symposium on Purum (Chote) Revisited.* Calcutta: Anthropological survey of India.

Livingstone, B. Frank. 1959. 'Letters to the editor: A further analysis of Purum social structure', in *American Anthropologist.* Vol. 61. 1084-87.

Needham, Rodney. 1958. 'A structural analysis of Purum society', in *American Anthropologist.* Vol. 60. 75-101.

Needham, Rodney. 1960. 'Brief Communications; Structure and change in asymmetric alliance: Comment on Livingstone's further analysis of Purum society', in *American Anthropologist.* Vol. 2. 499-503.

Needham, Rodney. 1960. 'Chawte social structure', in *American Anthropologist.* Vol. 62. 236-253.

Needham, Rodney. 1964. 'Explanatory notes on prescriptive alliance and the Purum', *American Anthropologist.* Vol. 66. 1377-1386.

Sangma, S. Milton. 1987. *History of the American Baptist Mission in North-east India.* Vol. 1. New Delhi: K.M. Mittal Publications.

Shakespeare, John. 1912. *The Lushai-Kuki clans.* Delhi: Suman Lata, Cultural Publishing House.

Shaw, William. 1929. *The Thadou-Kuki.* Assam: Published on behalf of the Government of Assam.

Singh, Wahangbam Ibohal. 1986. *The history of Manipur (An early period).* Imphal: Manipur Commercial Co.

Singh, K. S. 1994. *The scheduled tribes: People of India.* Vol.III. Delhi: Oxford University Press.

Yuhlung, Cheithou Charles. 2007. 'Matrilateral Cross-Cousin marriage among the Chothe of Manipur', in *Sociological Bulletin.* Jan-April, Vol. 56 (1), 46-64.

About the Author

Dr. Cheithou Charles Yuhlung is a Lecturer in Women's College, Shillong – 03. He has worked as Research Assistant as an evaluation team for IWGIA an international donor to assess the Naga Women's Union Manipur (NWUM) an NGO in 2007. He also worked as Research Associate in Dept. of Philosophy, under UPE-Project in North-eastern Hill University (NEHU) from 2011-12. He had a publication to his credit in Sociological Bulletin (2007). He also served as part-time Assistant Professor in William Carey University, Shillong, in 2012-13.

E-mail: Charlesyuh@gmail.com   Phone: 9862552550

Chapter 5

Sociolinguistics and Typological Aspects of Rongmei

☆ *Guigongpou Gonmei*

ABSTRACT

Rongmei is a language belonging to the Tibeto-Burman genealogical tree and are found spoken in the three states of Assam, Manipur and Nagaland. However, though the population is comparatively large, there are hardly any substantial works done on this language till date. Moreover, there are different dialectal variations within the language because of the geographical locations which in reality makes it more difficult for the researchers to fully bring out concrete works. This chapter is an attempt to present the Rongmei people and its Sociolinguistic aspects in the present scenario with special reference to Rongmei of Manipur.

5.1 Typology as a Discipline

Traditionally, typology was used as an alternative method in pursuing one of the same goals as generative grammar: to determine the limits of possible human languages and, thereby, to contribute to a universal theory of grammar. The paradigm result was the absolute universal law that would rule out as linguistically impossible what would seem logically imaginable, *e.g.* a language with a gender distinction exclusively in the first person singular.

Over the past decade, typology has begun to emancipate itself from this goal and to turn from a method into a full-fledged discipline, with its own research agenda, its own theories, its own problems. What has reached centre-stage is a fresh appreciation of linguistic diversity in its own right, and the new goal of typology is the development of theories that explain why linguistic diversity is the way it is - a goal first made explicit by Nichols's call for a science of population typology, parallel to population biology (1992:34).[1]Instead of asking "what's possible?" more and more typologists ask "what's where why?" Asking "what's where" targets universal preferences as much as geographical or genealogical skewing's, and results in probabilistic theories stated over sampled distributions. Asking "why" is based on the premises that (a) typological distributions are historically grown and (b) that they are interrelated with other distributions. Understanding distributions as historically grown goes back at least to Greenberg's (1965; 1978:122) and Givón's (1979:68-72) early calls for diachrony in typology and means that synchronic distributions, whether universal preferences or geographical clustering are seen as the product of type transitions and diachronic processes in general.

Understanding typological distributions as interrelated with and partly grounded in other distributions reflects the finding that linguistic structures tend to be systematically interrelated among themselves and with other anthropological patterns. Some of these findings gave rise to theories that predict close correlations between universal preferences in structure with universal preferences in cognition and communication (*e.g.*, processing preferences, as most extensively argued for by Hawkins (2004), and these have been at the top of typology's agenda. And last but not least, most typological distributions revealed distinct geographical patterns, and these can only be understood against models of population movements and language contact, systematically informed by what is known from population genetics and archaeology (see Nichols 1992, 1997; Fortescue 1998; Bickel and Nichols 2005; Dunn 2005).

However, pervasive reality effect makes clear that many current typological distributions can only be understood as the result of actual pre-history, both local and global. In turn, typological distributions provide a plethora of historical signals waiting for exploration and comparison with findings from other anthropological and historical disciplines.

5.2 Etymology of Rongmei

Rongmei is one of the sub-tribes of the Naga and is consider as one of the biggest among the Naga in Manipur (quoting as one of the biggest sub-tribes since it comprises the three states of Assam, Manipur and Nagaland, whereby the Rongmei Population according to the census of the three states put together stood at 2, 20,000 approximately).[2]

Rongmei means *'Men from the South'*. They are also known as *'Kabui'* or *Maruangmai, Inruanghmai, etc*. Rongmei is an ethno-linguistics term that stands for both the tribe and the language. The Rongmei population has been divided into a number of exogamous clans: *Kamei/ Kahmei, Gonmei/ Golmei, Gangmei, Rongmei/ Longmei, Dangmei, Panmei and Riamei, etc*.

5.3 Genetic Affiliation

Rongmei shares a close language affinity with languages like the Liangmai and Zeme which give rise to similar patterns of language usage and lexical arrangements though they may not be literally intelligible with each other. Rongmei also shares linguistic and cultural proximity with different language group such as Mao, Angami, Poumai, Thangal, Chakhesang, Kuki, Manipuri, Bengali (sylheti) etc., and due to these reason Rongmei has lots of borrowed elements.

Linguistically, Rongmei belong to the Tibeto-Burman family *(Sino-Tibetan)* according to Grierson (*Linguistic Survey of India Vol. III part II*, 1903).[3] But no proper classification has been given about them. But Rongmei in the name of Kabui falls under the Naga sub-section of the *Naga-Bodo* section under the Assam Burmese group of the Tibeto-Burman branch of Sino-Tibetan or Tibeto-Chinese speech family. Ethnically, they are of mongoloid origin and is believed to have migrated from the western part of Tibet along with other Mongoloid groups to the region.

5.4 Demographic Distribution

Rongmei are spread in three states: Manipur, Assam and Nagaland. In Manipur, they are found in the Tamenglong district and in the Imphal valley. In Nagaland, they are located at Peren district and at isolated places in Dimapur and Kohima. In Assam they are found in Cachar district around the Barak river valley of Silchar.

There is no accurate documentation regarding the population of the Rongmei. The recorded population of the rongmei is still in controversy. According to the 1991 census, Rongmei has a population of about 68,925(Manipur) where the male's population stood at 34,797 and females at 34,128. There are more than 195 villages. Presently an unofficial population is estimated at around 1,80,000 – 2,00,000 from all parts of India, with 60,000 from Manipur state alone. However, 2011 Census indicates a total population of 140,143 of Rongmei speakers including 94,758 of Kabui speakers.

5.5 Socio-cultural Aspects of Rongmei

Like any other Naga tribes, Rongmei has a rich culture and tradition which form an integral part of each Rongmei individuals. In their custom it is considered as a must that the bridegroom pays for the bride price in kind in the forms of bulls, land and other items, etc. So also the traditional dances of varied forms and style are still intact and practiced among the Rongmei till today. Festivals of different occasions and celebrations such as *Napthan-Ingai* (Harvest), *Gaan-Ingai* (New Year), etc., are still very much part of their culture. Traditional practices such as head hunting, barter system and so on has been stop with the modernization.

Rongmei followed the tribal form of worship like Nature worship, belief in the supernatural and animism. Taboos, curse and superstitions are also part of their practices. But with the coming of Christianity all these forms of practices has taken a back seat.

Rongmei are rich in their cultural attires, especially of women. The traditional dresses of the rongmei are very colorful and have wide varieties which represent the fun loving lifestyle of the people. The women wore a traditional dress call *'Pheisuai'* and *'Bungkam'* having different colours in it with varieties of motifs who wore them on different occasions like festivals, dances and mourning time. Other items such as *'Into/Toh'* (bangles), *'Beih'* (kind of necklace), *'Tobian'* (arm bangles), and *'Pikhim/ Pikam'* (head decorations) are also used by the Rongmei womenfolk. The dresses of the menfolk are very simple *i.e.* they wore a piece of cloth called *'Inthinoi/Thinoih'* to cover their private parts and they hardly have ornaments except for earrings and feathers attached in the head. However, all these forms of dressing have stop because they also have adapted to modern dressing code.

5.6 Literacy

Educationally, Rongmei are still lagging behind. Only about 50-70 per cent of the populations are literate where the males outgrow the females in terms of literacy percentage. Moreover, Rongmei is still a language, where linguistic work based on scientific and modern techniques has not been done to improve or to bring awareness to the general masses. In educational institutions and other levels of language development platform, they are way behind other dominant languages since no proper planning for the development of the language is initiated by the Manipur Government. Majority of Rongmei populations use *'Rongmei'* in their conversation among themselves, however, with other ethnic tribes or people they use either - Manipuri, Hindi or English. Lack of awareness among the people is also one reason why Rongmei language is still under-developed. Facilities to develop this language are not provided. However, few local platforms are helping in carrying this language forward such as in Magazines, pamphlets, Bible, Radio programs, newsletter, etc. Some of the books that have been published are story books, pamphlet, journals and history of the Rongmei people. Nothing on grammar or linguistic analysis of the language has been carried out. Therefore, Rongmei can be term as a handicapped language in many areas.

5.7 Brief Phonology of Rongmei

Phonology is the scientific study of sound system and its analysis and phonetic is the actual representation of speech sound.

Phonemic Inventory of Rongmei

Rongmei like any other Tibeto-Burman language is quite rich in phonemic inventory.

Rongmei phonology consists of 7 vowels (*a, i, e. o, u, ∂,c*) and 20 consonantal sound (*b, cʰ, d, g, j, k, kʰ, l, m, n, K▯, p, pʰ, r, s, t, tʰ, w, z, A▯.*).

Vowels

Vowels are sounds which are produced without the constriction in the air passage and the air flows out of the mouth freely and the sounds are relatively loud and strong.

Vowels Chart of Rongmei

Front	Central	Back
i		u
e		o
	∂,	c
a		

The vowels chart in Rongmei is clearly marked. Vowels in Rongmei are categories in the following labels namely:

1. High, 2. Higher- mid, 3. mid, 4. lower- mid, 5. Low.

Vowels in Rongmei occur in almost all the position except a few restrictions which are the phonemic contrast and the allophonic distribution.

The Consonantal System

In Rongmei, the consonantal system makes use of the distinction between aspirated and unaspirated, voiced and voiceless in the case of stops and fricatives. The other manner of articulation in which consonants are found to occur are nasal, affricative, trill, lateral and semi-vowel. Rongmei has 20 consonantal sounds. They are; *b, cʰ, d, g, k, kʰ, l, m, n, p, pʰ, t, tʰ, h, r, s, K⍰, j, y, z, A⍰*. They are found to occur in six places of articulation- bilabial, dental, alveolar, palatal, velar and glottal.

Tones in Rongmei

Tones are of two types namely; Register and contour tones. Rongmei has only register tones. Register tones are: high, mid and low; and Contour tones are: falling-rising, rising-falling.

Examples of Phonemic Tone in Rongmei

(high)	'si'	dog
(Mid)	'si'	bad
(Low)	'si'	sand

5.8 Gender Systems in Rongmei

Gender classifications in rongmei are very simple. Gender is not grammatically marked in the language, *i.e.*, gender is determined on the natural recognition of sex. Hence it has only natural gender. All the gender markers are suffixed to the noun root. So, we can say that the gender distinction in Rongmei is natural. That is, all the male beings come under masculine and all the female beings come under feminine. On the basis of semantico-morphological criteria, nouns are primarily classified into two classes, *viz.* i) animate and ii) inanimate. All the inanimate things are considered as neuter gender. Those animate beings which are not illustrated for sex fall under the common gender. Both the human and non-human nouns are differentiated for masculine and feminine gender. Nouns are assigned to a gender class on a strictly semantic basis. The gender of a noun is only determined by its meaning. The basic semantic distinction between the two genders is animate and inanimate. The animate nouns in Rongmei are morphologically marked for masculine and feminine genders. There are different gender markers for the human and non-human nouns. All inanimate things fall under the neuter gender. On the basis of the morphological marking the Rongmei Gender may be classified as in the following diagrams:

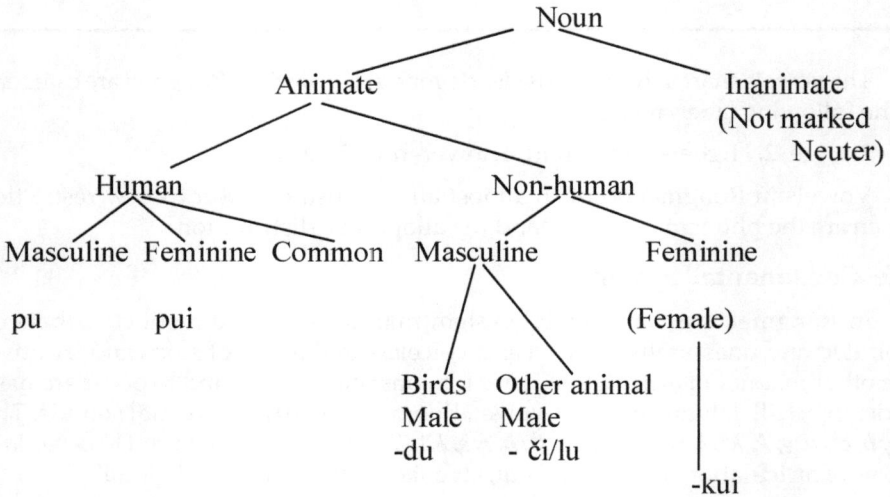

Chart 1: Morphological Marking of Rongmei Gender

Neuter Gender

Nouns representing in-animate objects. The gender is lexical.

dui 'water'

bi 'clay'

tʰiŋbaŋ 'tree'

In Non-human animate beings other than birds and animals. The gender distinction is never made: kho 'fish'

čakheŋ 'mosquito'

simaŋ 'fly'

Many human nouns not representing to professions belong to common gender.

əlau 'child'

puipu 'parents'

tuna 'girl'

The human animate masculinity is expressed by the suffix -pu

> taimai-pu 'manipuri man'
>
> tazuaŋmai-pu 'bengali man'

In attributive noun

> mbomai-pu 'mad man'
>
> talemai-pu 'flirty man'

The non-human animate masculinity of birds is expressed by the suffix –du/dui.

> ruai-dui 'cock'
>
> ruai-du 'chicken (male)
>
> phum-du 'duck (male)'
>
> phina-du 'pigeon'

The non-human animate masculinity of animal is expressed by the suffix –či /lu in most of the cases:

> miaunə-či 'male cat'
>
> guai-či 'bull'
>
> si-lu 'dog'

5.9 Some Basics Typological Characteristics of Rongmei

1. Rongmei follows the word order *SOV.*
2. Rongmei is *not a pro-drop* language.
3. Rongmei is *post positional.*
4. In Rongmei *adjectives follows the head noun* (in certain cases it is vice-versa).
5. *Genitive follows* the head noun.
6. Time adverbial precedes place adverbial.
7. Rongmei language has a close link between morphology and phonology.
8. *Contour tones* are considered by others as high and low register tones, however rongmei does not have contour tones.
9. There is *no tense* in rongmei, but it has aspectual markers. Aspects are used according to the context and time.
10. *'Mai'* in Rongmei are subordinate modifiers for nouns, verbs, and adjectives.
11. *'Bam'* are used for present progressive and *'khuan'* for past progressive.
12. Ergative occurs if the verb is transitive.
13. In rongmei there are no passive constructions.
14. When *–s* occurs with noun its plural but with verb it is 3rd person.

15. *Mai'* refers to nominaliser or adjectiviser, subordinate modifier (derivative suffix).

16. Meteorological verbs are those which refer to the weather verbs. Such as 'it is hot' *Will, might, should* are all modals (aspects).Rongmei exhibits these weather verbs.

17. In any given sentence in rongmei, there are the *participants, events,* and participants can be both object and subject. Then there is the background information.

5.10 Dialectal Variation among Rongmei

The number of dialect spoken by Rongmei is still not very clear. However, according to various views and opinions based on usage three varieties of Rongmei dialects are identified that closely resembles to each other. They are: the western varieties (Tamenglong), the south western valley (Khoupum & Nungba) and the eastern varieties (Imphal valley).

These dialects assigned different roles or functions depending on the domain they inhabit in the past. One factor is the inter-marriages with other ethnic groups and the geographical location. A Rongmei can speak all the three dialects in a given situation. But the southern and the eastern group exhibits more proximity.

Examples for these Dialects can be seen as given under

Western	tabui-tun	'where to'
South western	khou-tho	'where'
Eastern	thoukai-tho	'where to'
Western	guok	'pig'
South Western	gak	'pig'
Eastern	gok	'pig'

5.11 Language Use and Domains

Edwards. J says that "it has been commonly found that when the mother tongue of a minority language remains dominant in communication within the ethnic group, it can be said that the mother tongue is being maintained. If only intergroup language shift occurs, the language situation within the community will evolve towards a form of stable bilingualism" (1992:6).[4]

Rongmei are soft spoken people. They are normally polite in their use of speech. The phonemic intonations of the vocabulary/ words can change the whole meanings of the sentence or the end product. Stylistics usage varies with gender, where the quantity is more among men rather than women. Women seem to retain much of the indigenous vocabularies and utilized the language much better. Education also plays an important role in language norms (norms of speaking) among the Rongmei. About 60-70 per cent of the Rongmeis are bilingual and they adapt to code switching and mixing of languages. The educated Rongmei tend to use more of Manipuri and English than the uneducated as they consider these languages to have a better prestige or status. However, the older generations continue to use more of their mother tongue in day-today speech. Use of English and Manipuri does not necessarily bring about any quality of speech styles, but it helped improves in the process of interactions among different communities outside their domain.

Chart 2: Rongmei Language Use in Each Domain

Domain	Interlocutors	Activities	Settings
Family	Parents, children, siblings, spouse etc.	Mostly Conversation	
Education	Classmates teachers	Conversation	Classroom
Market	Rongmei and Non-Rongmei speakers	Communication and interaction	
Work Place	Colleagues, clients etc	Conversation and interaction	
Religious Places	God, other worshippers	Praying, songs and others	Church
Government	Government officials		Village level, township level, country level

Fasold stated that the concept of domains was first proposed by Fishman as a way of looking at language choice (1984:183).[5] According to Fishman, 'domains are institutional context in which one language variety is more likely to be appropriate than another'. Domains are factors such as; locations, activities, and participants choice. Use of Rongmei language in the respective domains for communication purpose has a wider scope for better understanding of the language since each domain has its styles and patterns of usage. Educational institutions and other officials institutions has been viewed as a place for standard language usage while normal conversations among peer groups and other places are considered as casual usage.

5.12 Bilingualism and Multilingualism in Rongmei

Bilingualism involves many aspects such as origin, internal and external, identification, competence and functions (Skutnabb-Kangas 2000). Bilingualism is defined as 'the knowledge and skills acquired by individuals which enable them to use language along with their mother tongue' (Blair 1990:52; Baker 2001).[6] Like most tribal people of North-east India, Rongmei are bilinguals. Along with their mother tongue, they speak English, Manipuri or Hindi in certain cases. However, Rongmei of Manipur often used Meitei/ Manipuri language to communicate with other linguistic groups the reason being that Manipuri is the Langua-Francua of State. Another reason is that the Rongmei language is not being taught in the school as a medium of instruction or a subject. So, the utility of their language is very less in comparison to the other languages like English, Hindi and Manipuri, etc. It may be claimed that the percentage of bilingual speakers among the Rongmei is 60-70 per cent . However, bilingual speakers are increasing particularly among younger generations in recent years due to the impact of media, such as radio, television, cinema, etc. in both the rural and urban areas leading to the borrowing of lexical items among Rongmei speakers. They loan words liberally from dominant languages: Manipuri, Hindi or English occasionally in place of their indigenous lexical items. Moreover, Rongmei vocabularies are not fully used by the younger generation in their speech which leads to mixed and lost of the originality of the language.

Rongmei is mostly monolingual, however, bilingualism and multilingualism does have a place in the Rongmei language scenario. Rongmei due to its close proximity with other neighboring languages like Manipuri, Angami, Assamese and Hindi, etc, borrowed and shared features similar to these languages. Moreover, they tend to learn these languages and started using as a means of communication for variety of purposes in their everyday life. A Bilingual/ multilingual Rongmei speaker will tend to adapt easier than a monolingual speaker. The number of languages spoken by Rongmei depends on the individual. These languages are used and assigned different roles or functions depending on the domains and situation they are in. Therefore, language use situation among Rongmei is different from what obtains in other bilingual and monolingual countries like Canada where there are two languages: French and English, and Japan where a single language is used for all activities.

Chart 3: General Usage of Rongmei Language by the Rongmei Speakers

Speaker of Rongmei	Monolingual	Bilingual	Multilingual
Village or Rural level	60 per cent	30 per cent	10 per cent
Town or Urban level	10 per cent	50 per cent	40 per cent
City Level	0 per cent	50 per cent	50 per cent

5.13 Language Maintenance among Rongmei

One of the basic underlying factors in studies on language maintenance and language shift is to discover the use pattern of individuals in a given community. Therefore, the study of language use has been the focus of many language maintenance studies. Fasold (1984), "Language maintenance is a sociolinguistic factor when a speech community collectively decides to continue to use the language that they traditionally used." Language maintenance is not only crucial but also a challenging task for each and every community to preserve to the extent possible its distinct linguistic and ethnic identity. However, for a minority community like Rongmei, home seems to be the main source for language maintenance. According to Anne Pauwels, "The term *language maintenance* is used to describe a situation in which a speaker, a group of speakers, or a speech community continue to use their language in some or all spheres of life despite competition with the dominant or majority language to become the main/sole language in these spheres". Fishman (1972) has described two different approaches in studying a language in the home domain.[7] One's "family" includes father, mother, children, domestics, and so on (Braunshausen and Mackey: 1962, 1965, 1966), as well as what Gross (1951) specified about grandparents *e.g.*, grandfather to grandmother, *i.e.*, the language of the interaction between speaker and hearer within the home domain. Undoubtedly, all Rongmei reported that they speak Rongmei at home and encouraged their children to use the language in their home domain. It may be the reason that they have a positive attitude towards their language and used it as a tool for identifying themselves as a distinct ethnic community. However, Rongmei use Manipuri and other dominant languages outside the home domain with non-Rongmei speakers. It is also obvious that institutional support is one of the factors which can empower an ethnic community to maintain its language by any speech community. But it is not happening in the case of Rongmei. That is, the Rongmei language is not taught in the schools or any other institutions so far, and yet their use of Rongmei in the home domain helps them maintain their language. However, lack of institutional support has restricted the lawful use of their language in many other domains. Hence, without institutional support they are able to maintain their language only in their home domain as the main source of maintaining language among the Rongmei.

Religion also plays a vital role in maintaining their language, because when they pray they use their mother tongue and the pastors, priest etc. either reads or recites the Bible or magical charms (mantras) too. The Rongmei has not changed much of their cultural values. They still wear their traditional dress and celebrate festivals with their songs and dances in the traditional ways. Every year Rongmei people celebrate the 'Gaan Ngai' festival in the month of December or January with joy and pride. Inter-ethnic marriages are not encouraged in the society but it occurs sometimes. They have a notion that if they allow Rongmei to marry persons from some other communities, ultimately they will be the losers, because they are less in number and the couples will no longer be able to maintain their language because they will find a common language for both, rather than continue the use of Rongmei and the traditions represented by it. However, the print media also helps one to maintain the language. There is a newspaper/newsletter called "Dicham" published locally and to some extend it helps the readers in learning rongmei language which in turns helps them to maintain their language to a certain extent. However, there are various social factors that affect language maintenance such as, status, demography and institutions. All these factors directly or indirectly affect language maintenance among the Rongmei speakers.

5.14 Language Attitudes

This connotation refers to people's feeling and preferences towards their own language and other speech varieties around them, and what value they place on those languages. Some language-attitudes studies are strictly limited to attitudes about language itself and some studies are broadened to include attitudes towards speakers of a particular language or dialect (Fasold 1984:148). According to Lambert (1967) attitudes consist of three components: the cognitive, affective and cognitive components (in Dittmar 1976: 181).[8] The cognitive component refers to an individual's belief structure, the affective to emotional reactions and the cognitive component comprehends the tendency to behave in a certain way towards the attitude (Gardner 1985). Language attitudes are the feelings people have about their own language or the languages of others (Crystal 1992). Our respondents were also asked a set of questions relating to the use of and their attitude towards their language, the responses to the questions indicate a clear picture that an overwhelming majority of mothers, children want to speak Rongmei at home which is closely linked to their linguistic identity.

Attitude towards Other Languages

When questions were asked to some select Rongmei what they feel on the use of and their attitude towards other languages such as Manipuri, Hindi and English languages. Their response shows a positive attitude towards other languages. They feel speaking these languages helps them develop their language skills, however, also feels threatened to some extent due to less political influences. Manipuri is used when they go to market, hospital and offices. Rongmei view English as one of the prestigious languages in the world, therefore they send their children to English medium schools rather than govt. schools. Moreover, the spread of Christianity have influenced their culture and created an interest and fascination for learning English.

5.14 Prestige and Status of Rongmei

Generally speaking, a language is considered prestigious if given prestigious functions and is considered low if it is not given any function at all or allocated low status functions. In actual sense, through status planning, the status of a low language can be considerably enhanced. For example, in Manipur, education is taken to be a high prestige domain of language functionality. And those languages that are used as mediums of instruction in educational institutions and government offices are given high or prestigious status; this is the case with English and Hindi. However, Rongmei which is not taught in schools or use in government domain cannot be classified as high and low variety. Moreover, the demerit is that there is no standardized dialect in Rongmei that can be categorized as having high prestige for official purposes, since such divisions among the dialects cannot be generalized. Every Rongmei dialect is based on areal locations only.

Rongmei Language Development

Development is used synonymously with modernization and standardization. The most basic measure of language development is graphization or representing in written forms which in turns can be used for references or taught. Other measures include availability of dictionaries and linguistic descriptions, lexical expansion, meta-language or register. All nationalities and people have the freedom to use and develop their own spoken and written language. And to preserve or reform in their own ways and customs some people are very pessimistic about the future of the minority languages because they feel their languages will disappear with the growing modernization. In Rongmei society, the functions allocated to a language seem directly proportional to the extent of their development. For instance, used of Rongmei language in Doordarshan, Media-broadcastings, Newspapers, etc. are because they have been in the stages of development. However, the language is yet to find it berth equal to the status of other MIL languages. This will need preparations and community participation, besides political backing.

Institutional Policies

Institutional policies of government ministries, agencies, cultural and religious organizations, language development centres, universities, other educational institutions and the media within the country contributes remarkably to the determination of language functions. Generally, languages that receive the institutional blessings tend to prosper functionally while those that do not tend to wither away. In Manipur, the case of development of language is different. Manipuri have the backing of the institutional policies of the government, however other minority languages including Rongmei are not given preference nor supported by any government institutions or agencies. Steps that are required to develop and enhance language are entangled in the politically motivated game.

Population Strength

The number of speakers of a language tends to contribute to its prestige and status and it affects its allotted functions. The state function allocated to Manipuri is traceable to the number of their speakers. Languages with a small population of speakers in Manipur are functionally underrated, oppressed and belittled. Certainly, the most noticeable aspect of language use situation in Manipur is the system of hierarchical distribution of functions among the different languages in the state. The indigenous languages of Manipur are the languages of ethnic solidarity and local day-to-day interactions. They are also predominant in religious worship and cultural programs of various kinds. It is significant to note that as far as functions of language in the Rongmei multilingual context is concerned, speakers modify their speech codes appropriately to fit changes of interlocutors, social setting, discourse topic, and conversational mood. The domains for the use of each of the languages in Manipur include the institution of government, the media, commerce, schools, religion and the home.

5.15 Historical and Political Profile

According to Adegbija (2004), to a large extent, the historical and political past tradition tends to attract greater functions to a language or languages. At the National levels, the Official Language is Hindi, Manipuri at the state level and at the international functions is the English, which enhanced by the political power-broken dynamism of the combined force of the native speakers. Hence, English language is allocated official functions in Manipur.

Socio-Political Development

Rongmei are politically divided in the three states of India. Benefits are reaped by the politicians for their political game plans, thus, affecting the language scenario to certain extent. In Manipur, Rongmei have little advantage in the political scenario, since the language is given less chance to incorporate in the functioning of the state language scenario.

Conclusion

This paper attempts to present a brief sociolinguistic aspect of Rongmei. Many things about this language remain to be done; the scope for the study of Rongmei language of North-east India is very broad. A scientific study and systematic analysis of the Rongmei language of North-east India may lead to new challenges. Detailed research can be conducted on the following points:

a) The descriptive study of Rongmei language,

b) Sociolinguistic study covering all aspects,

c) The historical and comparative study of Rongmei grammars.

Endnotes

1. Nichols, Johanna. 1992. *Linguistic Diversity in Space and Time*. University of Chicago Press.

2. http://en.wikipedia.org/wiki/Rongmei_Naga.

3. Grierson, G.A. 1904. *Linguistic Survey of India* (Vol-III). Government printing press, Calcutta.

4. Edwards, J. 1992. 'Socio-political aspects of language maintenance and loss: towards a typology of minority language situations', in W. Fase, K. Jaspaert, & S. Kroon (eds.). *Maintenance and Loss of Minority Languages*. Amsterdam: John Benjamins Publishing Co.

5. Fasold, Ralph. 1984. *The sociolinguistics of society*. Oxford: Basil Blackwell.

6. Baker C. 2001. 'Review of Tove Skutnabb-Kangas: Linguistic genocide in education – or worldwide diversity and human rights?' in *Journal of Sociolinguistics*. Vol.5:2. May 2001.

7. Joshua A. Fishman. Ed. Anwar S. Dil, 1972. *The sociology of language; an interdisciplinary social science*. Stanford: Stanford University Press.

8. Dittmar, Norbert. 1976. *A critical survey of sociolinguistics: Theory and application*. New York: St.Martin's Press.

References

Abbi, A. 2006. *Endangered Languages of the Andaman Islands*. Munchen, Lincom Europa.

Barua , P.N. Dutta and Bapui, V.L.T. 1996. *Hmar Grammar*. Mysore: Central Institute of Indian Languages.

Bhatt, D. N. S. and M. S. Ningomba. 1997. *Manipuri Grammar*. Munchen, Lincom Europa.

Chelliah, S. 1997. *A Grammar of Meithei* (Mouton Grammar Library 17). Berlin: Mouton de Gruyter.

Chelliah, S. and H. Th. Singh. 2007. 'The Lamkang language: Grammatical sketch, texts and lexicon', in *Linguistics of the Tibeto-Burman Area*. Vol. 30(1):1-212.

Chetan, U. 1976. *Structural Analysis of Mallipuri Language*. PhD Thesis, Manipur University.

Davies, Alan and Elder, Catherine. 2005. *The Hand book of Applied Linguistics*. UK: Blackwell publisher.

Deb Debajit. 2012. 'Bilingualism and Language Maintenance in Barak Valley, Assam: A Case Study on Rongmei', in *Language in India*. Volume 12: 1, January 2012 ISSN 1930-2940.

Dolen, H. 2004. *Structure of Manipuri (Meiteiron) Mealling*. Imphal: MALAPES International.

Fishman, J.1971. *Sociolinguistics: A brief introduction*. Rowley: Newbury House language series.

Fürer-Haimendorf, Christoph von. 1982. *Tribes of India: The Struggle for Survival*. Oxford: University of California Press.

Gonmei, M. 1994. *Introduction to the Rongmei Nagas*. Imphal: Yaiskul Janmasthan.

Greenberg, J. H. 1963. 'Some Universals of Grammar with Particular Reference to the Order of Meaningful Elements', in Greenberg, (Ed.), *Universals of Language*. Cambridge: MA, MlT Press, (73-113).

Imoba, S. and S. lbetombi Devi. 2004. *Manipuri to English Dictionary*. Manipur: Imphal.

Joseph, U. V. 2007. *Rabha*. Leiden, Brill.

Kabui, K. 1991. *History of Manipur*: Pre-colonial period. Vol.1. New Delhi: National Publishing House.

Kabui, G. 1997. *Jodonnang: A Mystic Naga Rebel*. Guwahati: SPN Associate Pvt. Ltd.

Kahmei, N. 1995. 'The Zeliangrong', in N Sanajaoba (ed.). *Manipur Past and Present*. New Delhi: Mittal Publication. Vol. III. (Pp.417,466)

Kurath, H. 1972. *Studies in Area Linguistics*. Bloomington: Indiana University Press.

Madhubala, P. 1979. *Manipuri Grammar*. Unpublished PhD Thesis, Manipur University.

Madhubala, P. 2002. *Mallipuri Phonology*. Manipur: Imphal. Potsangbam Bhuban Singh.

Ningomba, M. S. 1976. *Maring Grammar*. Unpublished PhD Thesis, Manipur University.

Ningomba, M. S. 1981. *Meiteilonmit* [Meitei Language]. Manipur: Imphal Board of Education.

Panmei. N. 2001. *The Trail from Makuilongdi. The Continuing Saga of the Zeliangrong People*. Tamenglong: Girronta Charitable Foundation Joyous guard.

Pamei, D. 1991. *Liberty to Captives – Z.B.C.C*. Tamenglong: Platinum Jubilee Publication.

Pamei, R.R.1996. *The Zeliangrong Nagas: A study of Tribal Christianity*. New Delhi: Uppal Publishing House.

Sharma, S. 2006. *Learner's Mallipuri-English Dictionary*. Manipur: Imphal, Sangam Book House.

About the Author

Dr. Guigongpou Gonmei had served as Research Assistant from 2010-11, and as Research Associate from July 2011-12 in UPE-project, in the Department of Linguistic, North-eastern Hill University (NEHU), Shillong- 793022. He had also served as a Guest Lecturer for a semester in 2012, in the same department. To his credit he has papers published in Dr. S.K. Singh (ed.). *2012. Ecology of Manipur,* and under UPE-project, The 'Sociolinguistics Aspects of Rongmei' in *Tibeto-Burman Linguistics of North-east India,* (Vol.3).

E-mail: ggonmei2009@gmail.com   Mobile: +91-7308963799

Section – II

Religion and its Practices

Chapter 6

Magico-Religious Belief of an Evil-Eye and the Traditional Curative Methods among the Tangkhul of Manipur

☆ *R.K. Jeermison*

ABSTRACT

Religion as an antidote in folk illness is well documented among the Tangkhul since time immemorial. Even before the advent of Christianity, the Tangkhul propitiated deities for bounty harvest and for good health. Rituals involving sacrificial of animal blood to please deities are historical. In recent times the social structure which is the outcome of 'belief' has significantly change due to Christianity brought about by Western missionaries. However, perspective on supernatural entities such as evil-eye continues and so as illness relationship to religion is sturdy.

Keywords: *Folk medicine, Religion, Evil-eye, Supernatural, Bio-medicine.*

6.1 Introduction

Much has written on folk medicine practiced in Africa and South America but rarely are accounted for India's North-east region where diverse ethnic groups inhabit and are currently defining the region's state of affairs. The region is noticeable with various ethnic groups competing for ownership of their only resources…, land and forest and the products, consequently creating conflict of ownership among them. The region can be classified as backwards or primitive because of its economy and direct dependence on the natural environment. The colonial imprint of social characteristics altering the traditions and belief system by introducing western education and various other variables, is indeed true, but has only led them to different direction. At best there is still the debate on what is modernity or civilisation. The world view is biased on this account. All that derived from the so-called civilized world of the west by Christian's missionaries and other agents employed by the colonial government is modern and civilized. On the contrary, the traditional science or technology and the ingenuity of the peasant community that addressed their lifestyles are considered backward and primitive. On the concept of deploying the term 'backward or primitive,' (seemingly based more on technology and civilization classification) there seems to be an invitation of debate that is most covered by politics. It is not fair to classify indigenous minorities or the peasant community or ethnic groups as backward and primitive, because many of them are more generous, pacific, intelligible and probably more honest from the moral point of view (Boedhihartono 1997). This is also true for the definition of 'Tribe' that cannot be universally defined or convinced as such that it evolved into the making of Scheduled Tribes based on certain criteria (exclusively for India). This terminologies (backward or primitive) deployed can be vaguely deduced as continuum of distance from the natural environment which they are dependent for their living. Ethnicity is also another categorisation that colonial ethnographers designate the Mongoloid stock that inhabits the forested tracts of North-east India. "It was during their rule that these identities assumed more solid forms as they became linked to territory and more disturbingly, history" (Bimol Akoijam 2012:3). Ethnicity is an inherent character of India's North-east which come in the way of its development and progress. The paper therefore attempts to place Tangkhul as a peasant community who are directly attached to the land and forest constructing their own culture in this region.

The Tangkhul is a Naga community living along the Indo-Burma border occupying Ukhrul district of Manipur, India and the Somra tract in Upper Burma. Ukhrul District is surrounded by Myanmar in the East, Chandel District in the South, Imphal East and Senapati Districts in the West, and Nagaland State in the North. The terrain of the district is hilly with varying heights of 913 m to 3114 m (MSL). The district Head Quarter- Ukhrul is linked with Imphal, the state capital, by NH-150 about 84 Km. The climate is of temperate nature with a minimum and maximum degrees of 3° C to 33° C. The average annual rainfall is 1,763.7 mm (1991). The exact location of the district in the globe is 24N - 25.41 N and 94 E - 94.47 E. The rainy season begins from May to October and winter is chilly. The terrain of the district is rippled with small ranges and striped by few rivers. According to Census of India 2011, Ukhrul district has a population of 183,115 with a density of 40 inhabitants per square kilometre (100/sq. mi.). Its decade 2001-2011 population growth rate is 30.07 percent. Ukhrul has a sex ratio of 948 females for every 1000 males, and a literacy rate of 81.87 percent (Jeermison 2012: 27).

6.2 Religion of the Past

The Tangkhul are believed to have migrated from Southeast Asia. They are now settled on the hills of Ukhrul in Manipur, India and the Somra tract of Myanmar. The construction of culture was mostly determined by its resources...land and forest. The forest provides food, timber, fuel and the list go on, and the land offer cultivation of agriculture crops such as rice, millet, maize, sesame etc. Making of culture is therefore very much visible to be determined by the resources mentioned. Shifting cultivation or *'Jhum'* is one important features that most of the traditional festive such as luira phanit (seed sowing festival) or the official *Lui-ngaini, Mangkhap phanit* etc. were constructed. Horam M (1988) writes... "all tribal festivities feature song and dance: the main tribal activity – agriculture has given rise to most of the songs and dances preserved till today. It is therefore, reasoned that forest and land provides the creation of socio-economic conditions among the Tangkhul substantially. When identified that "Culture" was constructed by the natural environment, there are various variables wherein culture was formed. One important variable *i.e.* traditional medicine (TM) has created "religion". This variable thus, determines the various norms and values that are prescribed to the community for the peaceful cohabitation.

Chronicles suggest that the real religion of the Tangkhul was that cult (or animism) which they professed to please annoyed deities. The belief is that there are numbers of deities assigned with varied portfolios - the forest deity, home deity, agriculture deity, rain deity etc. to be appeased so as to shun from their wrath at times. This predominantly animistic world view hold sways over the minds of various Tangkhul - a world view that sees man as a victim of nature or of fate. Going back to the past we see an "overwhelming animistic perception; that all natural events are viewed as ordered by the spirits, deities, gods and ancestors" (Ampadu: 265).

In the opinion of Horam, we see two distinct rites purely of Naga origin *viz.*; the family cult- the religion of the home, and the worship of the local divinity. He writes, of the two forms of ancestor worship mentioned, the former is the first in evolutionary order and the later being development. Among the Naga in the earlier times, there appeared to have been no difference between gods and ghosts, nor any ranking of gods as greater or lesser. He preferred the term ghost or gods instead of deities which we have employed earlier. Ancestor worship is simply to put as 'the ancient Naga thought of their dead as still inhibiting this world'.

Horam writes, "The ghosts of the departed were thought of as constant presences, needing propitiation by giving food and drinks. Each ghost must have shelter - a fitting tomb and offerings. While honourably sheltered and properly nourished the spirit is pleased, and will aid in maintaining the good fortune. But if refused the sepulchral home, the funeral rites, the offerings of food and drinks, the spirit will suffer from hunger and cold which will act malevolently and contrive misfortune" (1988:15,17).

A peep into the past reveals the people to be totally absorbed in the supernatural realm of the world. The Tangkhul were once labelled as animist that believes in multiple gods which are capricious, and unpredictable. Illness, famines and various social ills such as poverty and hunger are attributed to supernatural entities rendering the people to appease the deities. In semantics, appeasement and worshiping conveys different meaning which further complicated the definition of religion. "Emile Durkheim defined religion as, "a unified set of beliefs and practices relative to sacred things, that is to say, things set apart from forbidden beliefs and practices, which unite into one single moral community - all those who adhere to them" (Boedhihartono 1997).

6.3 Current Religious Positioning

Y. Milton said "Religion can be defined as a system of beliefs and practices by means of which a group of people struggles with these ultimate problems of human life" (1957:9). Before the advent of Christianity the people were strongly and strictly webbed to the natural environment wherein sacred trees and rocks are home to deities. The cultural displays to please the deities are in the form of sacrificial offerings. For any major event such as festival or rites, it starts with appeasement to the gods and deities to bring a bountiful harvest or to prevent the wrath of gods.

Religion and medicine are not viewed as separate concept or that aetiology of illness is explained away by religion. Like the 'Azande' of Pritchard and Gillies (1976), they were completely drowned into the pool of supernatural explanation for every un/wanted incidence. We therefore deduce that natural causes of illness are rare but are often supernatural.

With the advent of Christianity in the early part of 19th century brought about by colonial missionaries in almost every part of North-eastern region, the appearance of the social structure undergoes tremendous change. Animism, totemism, headhunting are other social evils targeted by Christian Missionaries. In its initial adventure the missionaries encountered several problems relating to cultural differences, but they were able to conquered the illiterate and savage society with much struggle. William Pettigrew was the Scottish Missionary that was deputed to work among the Tangkhul. Mission schools were established to absorb them into the western culture which is thought to be modern and civilized.

Appeasing gods or deities and ancestral worship were gradually replaced by Christianity. Various cultural practices such as head-hunting were eradicated. Most of these practices that come in the way of development or western civilization were surreptitiously erased retaining those practices which is deemed to be useful of colonial interest. The concept of religion and religious identities becomes more pronounced during that period. Cultural tradition such as festivals, agriculture (form of cultivation of crops) was fortunately retained but only in sync with Christianity.

6.4 Belief System

Since, the advent of modern medicine among the Tangkhul, numerous milestones have been recorded in the management of diseases that claimed to have save thousands of lives. As mentioned, the two phase of life, the evolution of religion from animism to Christianity of the Tangkhul indeed have restructured the society surfing towards Christianity, where belief system have change substantially. The belief that inspired culture, moulded worldviews and impacted the general development of Tangkhul in the area of hunger and poverty. In traditional societies "God is known everywhere as the omnipotent, omniscient and omnipresent Supreme Being" (Ampadu Chris: 266). The difference of the belief system of the past on medicine is the aetiology of illness subscribing to supernatural cause totally invalidating the concept of 'positivism' in social sciences. This mystic philosophy of the past have now rejoined to a more systematic approach. Identifying natural cause of illness as well as supernatural is now defining the aetiology of illness in the present set-up.

In olden days, the Tangkhul believed that the ancestral spirit, even though dead, continues to live among them. These ancestors are not worship but venerated (appeased). Animism here is the system of belief and practices based on the idea that objects and natural phenomena are inhabited by spirits or souls. Illness or medicine and religion cannot be therefore separated as this plays the cause and effect in the system.

Traditional religionists believed that diseases, barrenness and sudden death are all the works of witchcraft. They are not interested in scientific solution or medical attention since they believed their problems are caused by witches who must be appeased. In this context therefore, we see the position of religious healing over taking the relevance of modern medicine. Religious healers and diviners act as mediators between the real (man) and imaginary world (gods). "They understand the language of the spirits and therefore can foretell future events and happenings. They are the rain makers who can bring rain in times of draught, the sorcerers, witches and wizards who can cause pain, diseases and even death" (**Ampadu Chris**: 266).

6.5 Micro-Religious Orientation in the Aetiology of Illness and Healing Practices

A systemic approach to the study of supernatural elements is important to enable us to understand the complex relationship of religion and medicine. Empirical research provides ample evidence that religions shape their adherents' understanding of diseases and illnesses, their health related behaviour, their interactions with the expectations of the health care systems and their adherence to medical recommendations. Culture developed based on the influence of the natural or physical environment. And in this culture there is religion as a powerful determinant shaping the behaviour of a group of people or individual. Religion as a remedy/ medicine to illness is therefore well attested among the Tangkhul. Environment - Culture (religion, technology) - Behaviour (action and belief) forms a continuous cycle influencing one on the other. Environment subjective or objective is modified by technology which in turn altered the original and the first culture subsequently changing the behaviour of the society or the individual. Thus, religion is prone to alteration, subsequently changing the concept and definition at various times. As mentioned, religious is not static but is evolved and is still evolving. The influence of religion on medicine in traditional societies is a priority for choosing the medical plurality and its right healer. Supernatural entities are solely the cause of illness in the earlier phase of its existence. Calamities, diseases, famine are god centred framework in the interpretation of health and illness. However, today we see the difference of natural and supernatural causes of illness. What is "natural cause of illness" is clearly identified among the Tangkhul. Illness attributed to climate, food, accidents etc. are treated by herbalist, chiropractor, mid-wife, bone setter etc. and illness caused by supernatural elements is best treated by the diviner, exorcist and other religious healers believed to have certain power to communicate with the real and the imaginary world.

Religion therefore is an antidote to illness caused by supernatural entities. Allopathic practitioner is irrelevant in cases diagnosed as evil-eye, demonic possessions, sorcery, witchcraft and other godly illnesses. These types of illnesses are referred to religious healers and traditional healers such as the diviners and sorcerers. The Church and traditional healers (Shamans/ Khanongs) who are visibly different in the past are now working together. The people have converted to Baptist denomination and so the traditional local healers which were once secular are now becoming religious. Methods employed by the traditional healers to treat patients suffering from evil-eye, demonic possession etc. used religious tools *viz.*; Bible, Cross, Holy water and prayers and they also chant and performed rituals related to Christianity, instead of using the traditional magical charms. The principle of healing and curing various human ailments seems the same between the non-Christian and Christian believers but the difference lies to whom they address and prayed seeking for remedy.

6.6 Supernatural Aetiology of Illness

Various illnesses in Tangkhul society are explained by supernatural agents. Cases such as evil-eye *(rai)*, sorcery *(kharakasang)*, black magic *(laiva)*, witchcraft and demonic possession *(chipee kazang)* are very much common in the region dominated by the Tangkhul. There is no difference among population of various socio-economic background to believe in illness associated with supernatural explanation, and that is treated and cured by traditional and religious healers. The image of human body is interpreted holistically. Therefore, families opted for traditional or Christian healers in search of a holistic and complete cure. The intrinsic characters understand in various traditional societies is the belief of 'soul' in human body. This existence of soul or the spirit in the body explains much of the concept of health and illnesses for cultures such as the Tangkhul. We see that health and illness is the product of imbalances of the soul and the body, of the body and its physical environment, of the body, soul and the culture that defines traditional societies. These elements; soul, body, physical environment, culture are altogether related and should be balance or otherwise produce suffering in the form of illnesses, famine, dysfunction family, and other disorder and disharmony in social space. Health, therefore is not solely explained by illnesses or diseases as reasoned in biomedicine such as "a good health are those without disease", but is seen from the wholeness of the various natural and supernatural elements operating in cultures of such traditional societies. Therefore, causes of illness are not only explained by the natural variants but also by supernatural elements in the philosophy of Tangkhul culture.

The concept of health and illness for the Tangkhul are therefore a complicated function related to various cultural performances and not only subjected to the deteriorating health condition of the individual. In doing so, we see that health and illness cannot be separated from religion. Drawing boundary from one social group to others can be easily identified if health care system of a particular group is properly understood. What we meant are those cultural and physical factors that have subsequently defined the community's perception of illness. Every illness is more or less interpreted by a cause. The immediate causes of any illnesses are attributed to supernatural explanation and therefore, the first choice of treatment is traditional medicine. The cause brought about by evil spirit or annoyed deities, wrath from God, sorcery and evil eye initially defined the concept of health and illness in the community. But when this traditional knowledge and Medicine (TM) fails to response then only biomedicine became relevant. However, with the belief that illness is the subsequent form of violating norms and values the community embraced from time to time.

6.7 Evil-Eye

Evil-eye is the name for sickness transmitted - usually without intention - by someone who is envious, jealous or covetous. It is also called the invidious eye and the envious eye. This belief is that a person - otherwise not malefic in anyway - can harm people by looking at them with envy and praising them. Almost everywhere such Evil-eye is believed to exist. Its effects are said to occur as an inadvertent side-effect of envy or praise. Mentions of the evil eye (*ayin ha'ra*) in the Bible clearly refer to the role that envy and covetousness play in its development. We find in Proverbs 23:6, "*Eat thou not the bread of him that hath an evil eye, neither desire thou his dainty meats*", and likewise in Proverbs 28:22, "*He that hasteth to be rich hath an evil eye, and considereth not that poverty shall come upon him*".

In the New Testament book, we also see that when Jesus Christ lectured about defilement, he told his followers that *ayin ha'ra* comes forth from a man and defiles him just the same as if he had committed a physical crime: "From within, out of the heart of men, proceed evil thoughts, adulteries, fornications, murders, thefts, covetousness, wickedness, deceit, lasciviousness, an evil eye, blasphemy, pride, foolishness" (Mark 7:21-22).

Evil-eye is the consequence of envious eye and it involves without any intention. However "in Sicily and Southern Italy it is believed that some people can deliberately cast the Evil-eye on others. There the regionally idiosyncratic belief is that certain people (including at least one former Pope) are born with the Evil-eye and "project" it involuntarily. Such people are called jettatores (projectors) and their specific form of Evil-eye is called jettatura (projection) in contradistinction to the garden variety of envious or praising Evil-eye, which in Italian is called *mal occhio* (bad eye). Jettatores are not necessarily evil or envious people, according to this belief system, and they are often represented as being saddened and embarrassed by the harm they cause" (Bir Alev 1995).

The Evil-eye is an ailment that is very common among small children. It is believed that it is caused by excessive affection. If a woman or a man sees a child with physical attributes which he/ she admires, he/ she must touch the child and invoke God's protection so that the baby will not suffer from the Evil-eye. Children seem to be most susceptible to this ailment, although adults may suffer from it occasionally. Babies suffer the direst consequences.

6.8 Characteristic Relating to Evil-eye among the Tangkhul

1. In Tangkhul dialect Evil-eye is known as *rai* and the person who can cast an Evil-eye is considered as *phasa makathar* (*phasa*-body, *makathar*-unclean) and called *rai kaphung* (*rai*- evil spirit, *kaphung*- posses). *Rai* is an occult paranormal phenomenon where it attacks the weakest member of the family out of greed, jealousy and animosity.

2. Traditional healers who cast out Evil-eye are called *Khanong*. Christian healers such as pastors, clergies, deacons and syncretic churches also cast out these intangible forces by employing religious techniques. Although they believed Evil-eye and demonic possession as different entities but today they considered all possession as evil and therefore spiritual healings makes no difference.

3. *Khanongs* today are religious oriented and therefore in treatment both traditional practices (employing herbal and parts of animal) and religious procedures are followed.

4. It is believed that Evil-eye is usually hereditary. A family with such history will passed the power from generation to generation generally along the female line; from grandmother to mother and to daughter.

5. A male member also inherits this supernatural power but it is believed that the power is less effective as compared to females. Usually it remains more or less recessive. However, the power is passed to his spouse and daughters who become very active. Therefore, the possessors are usually women.

6. It is believed that there are variations in the power of Evil-eyes. The powerful can even fell a bird flying by merely gazing at it. They do not even stop from attacking her near and dear ones.

7. Casting an Evil-eye is without intention although jealousy, greed and envy could be one of its reasons. It is of the belief that even the individual who got such supernatural power are oblivious of his/ her deeds in the initial stage.

8. The community cannot openly discuss Evil-eye because they are scared that Evil-eye will possess them. With obvious reasons the people dare not inform the Evil-eyed person or the relatives about the supernatural power he/ she possess and that it is harming them.

9. A person with the Evil-eye will always posses this occult power throughout her lifetime. However, after the inception of Christian doctrine it is believed that one can denounce it except through deep devotion to Christian faith.

10. Evil-eye's patients are usually women and children. They are generally considered to be weaker because of its dependency, physique, and its sense of inferiority. They also usually fall prey to Evil-eye because they are more charming and decorative.

11. A patient after getting recovered might again get possessed by the same spirit or another else if the patient does not take precaution.

6.9 Traditional Healing Methods

A mother whose child had been struck by the Evil-eye will take the advice of the other women in her community and acquire an amulet for the child to wear to repel from the Evil-eye in future. This sort of charm is called a repellent talisman or *apotropaic* charm. There are preventive measures for averting Evil-eye among the Tangkhul as practiced in many cultures too. Some believed that although there are measures to prevent the cause, these are of little help. It is 'inevitable' as one informant suggest: 'No one can stop if the person possessing the Evil-eye is envious or jealous with another, certainly the harm is seen. But to put a cure to this can be successful depending on the power of the Evil-eye and the spiritual and physical strength of the victim'. Having confidence in one's religious faith or traditional and Christian healer is important for effective healing with a belief that they also have divine power, the ability to deal with the unseen mysterious forces and its ability to convey message to the spiritual world.

As mentioned, illness from Evil-eye means that the spirit of the person having evil power has overtaken completely the body of another identity and thus causing harm without intention. It is in this context of cross identity that healers diagnose Evil-eye. When a person is possessed, the behaviour, language, and other forms of action took the semblance of the person who is causing the harm. Symptoms diagnosing Evil-eye includes sudden stomach cramp followed by flatulence, diarrhoea, unusual behaviour like mental retardation, persisting headache and body ache, muscle contraction and seizures, high fever, a lack of appetite and sleep, and usually a swelling on some part of the body. The patient also usually falls into trance. Another form to diagnose the ailment includes placing the patient under the shade of *kurao angouba (Erythrina indica)* tree. If the patient is unwilling and is frightened to do so then it is diagnosed that the ailment is caused by Evil-eye. Inquisition of the patient history such as his/her whereabouts in the past two days, enquiring the patients' family of any guest visitation before the illness, and if any person has made an envious eye or appreciation of anything in the possession of the family are other forms of diagnostic procedures usually employed by traditional healers. Christian healers resorted to techniques of using the cross, placing the Bible on the patients palm and sprinkling holy water. One main basis for declaring a person possessed seems to be a violent revulsion toward sacred objects and texts. Christian healers hardly differentiate demonic possession and Evil-eye. For them both are the handiwork of Satan. Treatment of illness caused by Evil-eye often employs shock therapy and unsympathetic measures. Since that very person is over taken by the evil spirit and using his body, therefore therapies resorted are cruel in nature.

As mentioned, *kurao angouba* is used to diagnose illness caused by Evil-eye. The evil spirit is said to be scared of this tree because of its thorns and the corrugated bark that resemble numerous eyes (doubtful on the content of chemical compound). Therefore in treatment this tree is often used in the form of forcefully making the patient wear a carefully prepared talisman made of the bark or the wood from the tree. It is said that when the patient is made to wear, the evil spirit experience discomfort eventually giving up possessing it.

The other form of traditional therapy employed is the input of a "chilly" (*u-marok*) which is found in the area and have one of the hottest scoville heat units in the world. This is burned in a closed room where the patient rest. The obnoxious and choking scent produced from the burning chilly is intolerable for any human being. This makes the evil eye suffer and attempt in any way to resist leaving the patient. But with the repetition of this procedure, the evil spirit gives up the patient from possessing it.

A tropical root plant named *Curcuma caesia* or black turmeric and locally known as *yaimu* that is black in colour and extremely bitter is also used in the treatment caused by Evil-eye. The decoction of *yaimu* with incantations/prayers smeared it on the belly of the patient is another procedure narrated by one healer. The decoction is also prepared for drinking purposes. Since it is extremely bitter and astringent, an individual would experienced the phenomenon of discomfort. This taste, of course, paves the way for the disappearance of the Evil-eye. These are some of the methods employed by the Traditional healer to treat evil-eye and demonic possession, besides other methods that requires critical analysis.

Table 6.1: Treatment techniques for Evil-Eye Reported by Tangkhul Folk Healers

Vernacular; Meitei (M) and Tangkhul (T)	Scientific/English Name of the Antidote	Prepared form
Kurao angaoba, (M)	*Erythrina indica*	Wood or bark for locket/Amulet
Yaimu,(M)	*Curcumia caesia*	pasted and concoction for drinking
Umarok(M)/Sivathei (T)	Bhut jolokia* (ghost pepper)	Burned chilly. The astringent odour for inhaling
Haru (T)	Egg	Passing an egg over the body and then placing it in a bowl under the child's pillow overnight.
Hotla (T)	Ashes/Cinders	Powder the charcoal to ashes and mix it up with the water in the cup. Wait for few minutes to dissolve the ash and finally the ash will settle down producing a crystal clear potion.
Champara	Lemon	drinking

* The *Bhut Jolokia* is an interspecific hybrid cultivated in the Indian states of Nagaland and Assam. It grows in the Indian states of Assam, Nagaland and Manipur. There was initially some confusion and disagreement about whether the Bhut was a *Capsicum frutescens* or a *Capsicum chinense* pepper, but DNA tests showed it to be an interspecies hybrid, mostly *C. chinense* with some *C. frutescens* genes.

Illness aetiology on Evil-eye is not a thing of the past but is very much common today. However rituals which were prominent in the family and society in the past are buried with no significant replacement today except with Christian doctrine. As mentioned, this supernatural phenomenon is dealt from the perspective of religion. In the past and in Naga folk tales rituals involving nature/supernatural- land, animals, plants, spirits, deities etc. and its relationship to man's activities such as in food gathering, agriculture, illnesses, birth, death etc. are well documented. However after the inception of Christianity this has replaced completely by Christian doctrine.

Table 6.2: Some Illnesses Treated by Traditional Healer Using Medicinal Plants

Illnesses/Diseases English/ Local Name	Vernacular: Meitei (M) and Tangkhul (T)	Scientific/English Name of Herbs	Prepared form
Dog bite, cough and cold, Dysentery	Mangke, (M)	Tamarind seed.	Paste on the wound
Allergy	Khaimaithei (T)	Pumpkin	Boiled
Synusitis, itching and ring worm	Herbs	Drymaria cordata (L.) Willd. ex Schult.	Smeared paste
Constipation	Nongban lei (M)	Lantana/Camara Linn.	Juice
Regulation of blood pressure/diabetic	Chengkruk-tingkhan panbi (M)	Amaranthus/spinosus Linn.	Cook/eaten
Intestinal worm, cough, fever and for blood circulation	Kanghu or greater galangal (M)	Alpinia/galanga	Rhizome paste
Snake bite	lin-cheisoo (M)	Arisaema/tortuosum Schott.	Rhizome decoction
Epilepsy	Awa-phadigom (M)	Eryngo (E) Eryngium/foetidum Linn.	Vegetative/decoction
Skin disease and evil eye	kurao-anganba (M)	Coral tree (E) Erythrina/indica Lam.;	Bark powder and boiled liquid form
Skin disease	Kurao-angouba (M)	Erythrina/variegate Linn.	
Stomach ulcer	nong-gang-hei (M)	Euphoria/longana Linn.	Fruit
Ulcer, cough, dysentery/have socio-religious value	Khongnang-taroo (M),	Ficus/benjamina Linn. weeping fig (E)	Leaves eaten raw
Brain coolant and hysteria	kombi-rei (M)	Iris/bakeri walli	
Urinary stone case	Nungai-peruk (M)	Linaria/ramosissima Linn.	Soup
Jaundice and kidney infection	Lam ikaithabi (M)	Mimosa pudica Linn.	Soup
Dysentery, diarrhoea, haemorrage, annemia,	Heikru	Phyllanthus emblica Linn.	Fruit juice
Spleen, liver diseasesand abdominal complaints	Heirangkhoi	Amoora rohitka W & Arn.	Bark
Skin diseases	Yai-ngang	Curcuma caesia Roxb.	Rhizomes are aromatic tonic, stimulant, blood externally applied to sprains and wounds. ii. Fresh juice for skin affections.

Source: Field survey, 2010-2012.

Prayer has become the main antidote for any inconveniences that obstruct even livelihood. Apart from evil-eye which we have dug in, there are other illnesses worth to mention concerning the Tangkhul. The most commonly reported illness among the Tangkhul in this present set up is diabetics. This ailment is most common on adult population in the age group of 50 years and above. HIV / AIDS are more prevalent in the age group ranging from 25-40 years. Cases of tuberculosis and jaundice are not ruled out. Diarrhoea and pneumonia frequented children. The most dangerous scenario reported today in majority of the village is the abuse of drugs among the youths. Drug addiction has taken toll on large scale. Indeed the most important health issue to be addressed upon is on the consumption of narcotic substances. However, as our research is most concern on supernatural ailment especially on evil-eye, let us focus to the subject to avoid complexity.

Table 6.3: Some Medicinal Plants Used for Jaundice and Hepatitis-B Treatment

English	Botanical	Manipuri
Ceylon Leadwort	*Plumbago zeylanica* L.	Telhidak angouba
Gulancha	*Tinospora cordifolia*	Ningthoukhongli
Papaya	*Carica papaya* L.	Awathabi
Garden mint	*Mentha viridis*	Nungshihidak Pudina
Black Nightshade	*Solanum nigrum*	Leipungkhanga macha
Indian Gooseberry	*Emblica efficinalis*	Heikru /Amla
Phlogacanthus curviflorus Nees	*Nongmangkha ashinba*	

Source: Dr. K Paochunbou, Sangai express. 21st May 2013.

6.10 Discussion

According to Marcia Carteret (2011), "Folk illnesses tend to carry religious overtones as well as a range of symbolic meanings with social and psychological dimensions. A person suffering from folk illness is often seen as expressing emotional distress through the physical body. Such distress may arise from conflicts within the family, or from the larger social world that the individual inhabits. Perhaps the patient failed to observe social norms or perform essential rituals, or maybe his illness has been caused by an evil spirit".

The account has given insightful meaning of how culture embraces an important place in the explanation of illness aetiologies. Belief, attitude, taste, behaviour are to large extend moulded by the cultural environment in which the individual is brought up. Events caused by supernatural entities make no difference for educated and uneducated people. This is uncovered through discussion initiated from teachers and illiterate traditional healers. In traditional societies, the conditions and treatments of health and disease are influence by a combination of cultural, biological and environmental factors. Ethno-medical studies have established that particular populations have their own system for identifying the causes of illness, their own diagnostic procedures and therapy (Yoder 1989: 45). Social and cultural factors are so strong in traditional societies that these determine illness aetiology and their distribution. Seeking health care therefore became the product of various socio-cultural indices. The fundamental beliefs causing illness in many primitive societies are believed to be from sorcery, evil eye, breach of taboo, spirit intrusion or soul loss (Clement 1932: 185-252). But in modern medicine these cause of illness are the concern of irrationality, unscientific, and fraught with obscurity. "Many of the signs and symptoms experienced from evil-eye and demonic possession can be explained away by modern medical science. Seizures and convulsions are symptoms of epilepsy. Personality changes can indicate hysteria, or schizophrenia, or other psychological malfunctions. Lewd and obscene acts can indicate mental disorder. Having sexual thoughts, if taken seriously as a sign of demonic possession, would indicate nearly all of the modern population is possessed, especially the men. Distended stomachs can indicate malnutrition and other medical disorders. Also, having knowledge of future events or information is known as clairvoyance by many occultists and Neo-pagan witches which they consider a special spiritual gift. In light of such evidence it seems demonic possession is hardly functional anymore. But what we find in many traditional societies is that traditional healers indeed put a cure to these illnesses without employing allopathic drugs. Merely applying herbal medicine or by incantations and displaying sacred objects to the patient have cured illnesses.

Now the questions confronting us here are how a person suffering from epilepsy (if that is the case for science) be cured merely by incantations and elegant spells? A study of Roman Catholic in Sri Lanka reported that demons attack a weaker member of the chosen victim's household. Thus, women and children, who are considered weaker than men, are more liable to demonic possession, even though the actual sorcery is frequently directed against the men of the house (Stirrat 1977:133,157). The account is similar for the Tangkhul. Therefore, if children and women who are considered generally weaker are the ones that suffer the brunt of evil eye and demonic possessions, then we are further perplexed when the signs and symptoms are translated to biomedical diagnosis. Epilepsy of course is common in all age groups, but hysteria and schizophrenia are uncommon among children. We also know that schizophrenia strikes in similar rates among men and women in all ethnic groups. Then we can ask whether it is a fortuitous incident among many traditional societies that schizophrenia and hysteria frequented children and women.

Evil-eye has a middle-eastern, Mediterranean, and Indo-European pattern of distribution and was unknown in the Americas, Asia, Sub-Saharan Africa or Australia until the introduction of European culture. Dundes theorized that evil-eye is based upon the beliefs about water equating to life and dryness equating to death. The harm caused by overlooking consists of sudden vomiting or diarrhoea among children, drying up of milk among lactating mothers or livestock, withering of fruits on orchard trees, and loss of potency among men. In short, the envious eye "dries up liquids", a fact that he contends demonstrates its Middle Eastern desert origins (Dundes 1981). In Dundes theory we observed elements of dry and wet, and in Latin America there is a principle of hot and cold theory. Evil-eye (*mal de ojo*) is equated to hot illness caused by a strong or hot vision. Among the Tangkhul we found the concept of evil eye hardly relating to hot-cold and wet-dry principle, although hot or dry by and large resembles the symptoms. Evil eye is the possession of the evil spirit of another person and therefore treatment follows harsh and unsympathetic measures.

Various traditional societies are of the belief that supernatural entities do exist and is one of the major causes of illnesses. A study by Behura put forward that the Saora have their own concept relating to the aetiology of illness, believing that sickness is caused by the wrath of *kittung* (gods and goddesses), by spirit intrusion, by sorcery, by casting Evil-eye and by the breach of taboos. Diseases such as leprosy, smallpox are a condition being visited by god or goddesses because of breaching the taboos (Basu 1990:275-86; Behura 2003:27-67). A study of Khairwars in Madhya Pradesh revealed that some of them blamed it to the attacks by ghosts and evil spirits as the reasons for falling sick (Pandey 1996). Diseases caused from this divine wrath are the beliefs of many societies (Hockings 1980; Behura 2003:27, 67). Since it is this belief, the only therapy a person suffering from supernatural illness is to propitiate the god and goddesses. Therefore shamans, priests and exorcists confer a respectable status. Although the subject circumvents scientific rationality, it is the beliefs that concern a society and this cultural mode of understanding drives the people to avail health care from traditional folk healers even today.

6.11 Conclusion

We now know that supernatural explanation of illness aetiologies are confined not only to the Tangkhul but with various other societies. Explanations of supernatural events do vary from society to society. People are aware of the natural causes of illnesses; they know that bad food can entail stomach cramp and traditional herbal medicine or allopathic drug can stop the pain. But what they also know is that in some cases, drugs and doctors are ineffective, that doctors are unable to treat some diseases. In those cases they do explain, by reference to the supernatural entity (Deliege 2007:53). Tangkhul also never put down every illness to supernatural; they are very much aware of natural causes of illness and thus seek health care from allopath or traditional herbalist. There has been a paradigm change after Christianity became the main religion. In the past when Tangkhul were animistic, Evil-eye and possession from annoyed deity were clearly classified, but in the contemporary era these classifications have become very obscure. Although traditional healers, Christian healers and the general population believe in the Evil-eye, but for Christian healers these phenomena is interpreted according to the Bible and is considered both Evil-eye and spirit (deities) possession as demonic and bad.

References

Akoijam Bimol. 2012. *Remembering U A Shimray*. A seminar paper presented on 2nd Memorial Event at Tribal Institute, Imphal, 28 July 2012.

Ampadu Chris. Available at *www.wciu.edu/docs/general/ampadu_article.pdf*ý

Basu, A. M. 1990. 'Cultural influences on health care use: Two regional groups in India', in *Studies in Family Planning. Vol.* 21 (5): 275-86.

Behura, N. K. 2003. 'Health Culture, Ethno-medicine and Modern Medical Services' in *Journal of Anthropological Survey,* India. Vol. 52: 27-67.

Bir Alev. 1995. '*Evil Eyes' in* http://www.luckymojo.com/evileye.html

Boedhihartono. 1997. 'Local Religion and Traditional Healing Practice. The indigenous minority groups of Indonesia'. Indira Gandhi National Centre for the Arts, New Delhi. Available at htpp://ignca.nic.in/cd_05009.htm.

Clements, F. E. 1932. 'Primitive Concepts of Disease', in *American Archaeology and Ethnology.* University of California Publications. Vol. 32: 185-252.

Deliége, R. 2007. 'Demonic Possession in Catholic South India' in *Indian Anthropologist. Vol.* 37 (1): 49-66.

Dundes, Alan (ed. 1981.). *The Evil Eye: A Folklore Casebook.* New York: Garland Publishing, Inc.

Ellen, R. G. 1989. *The Encyclopaedia of Witches and Witchcraft,* New York: Fact on Files.

Hockings, P. 1980. *Sex and Disease in a Mountain Community.* New Delhi: Vikas Publishing House Pvt. Ltd.

Horam M.1988. *Nagas old ways and trends.* New-Delhi: Cosmo Publications.

Jeermison R.K. 2012. 'Perception of Health care option and therapy seeking behaviour of the Tangkhul Nagas' in *Journal of North-east India Studies*. Vol. 2(1), 2012.

Marcia Carteret. 2011. *'Folk Illnesses and Remedies in Latino Communities' in* http://www.dimensionsofculture.com/2010/10/folk-illnesses-and-remedies-in-latino-communities/

Pandey *et al.,* 1996. 'Socio-cultural Characteristics and Health seeking Behaviour of the Khairwars of Madhya Pradesh' in *The Journal of Family Welfare.* Vol. 42 (2).

Pritchard E. and Gillies E. (1976) *Witchcraft, Oracles and Magic among the Azande.* Oxford University Press: USA (Abridge edition).

Stirrat, R. L. 1977. 'Demonic possession in Roman Catholic Sri Lanka' in J*ournal of Anthropological Research. Vol.* 33: 133-57.

Tierney, E. O. 1984. *Illness and Culture in Contemporary Japan: An Anthropological view.* Cambridge: Cambridge University Press.

Yinger Milton .1957. *Religion, Society and the Individual.* New-York: The Macmillan Company.

Yoder, Y. S. 1991. 'Cultural Conceptions of Illness and Measurement of Changes in Morbidity' in Cleland and Allan, G. Hills (eds.). *The Health Transition: Methods and Measures: The Proceedings of an International Workshop.* London (1989) Health Transition Series-III, the Australian National University. pp: 43-59.

About the Author

R.K. Jeermison is an Asst. Professor in Department of Geography, Oriental College Imphal, Manipur.

E-mail: jeermi07@gmail.com

Chapter 7

Religious Reform and Sacralisation of Space: A Study of Heraka Movement in North-East India

☆ *Soihiamlung Dangmei*

ABSTRACT

The Heraka is a socio-religious reform movement of the Zeliangrong Naga derived from their traditional religion. It was organised from disparate groups of the early 1930s into a centralised and effective movement in 1974 in Assam, Nagaland and Manipur. Initially the movement was started by Jadonang in 1929. But due to his early execution, the movement was carried forward by his disciple Gaidinliu. The reform aims in the abolition of the obscurantist customs and superstitious practices. Heraka means pure, which is not mixing with other evil things. There are scholars who argued that the Heraka is a debased form of Hinduism, while others tries to maintain the Heraka distinct religious identity. Forging of religious and cultural identity based on indigenous tradition forms an important aspect of the Heraka movement. This chapter is an attempt to analyse the Heraka religious reform, particularly the sacralisation of space, and the institutionalisation of priesthood.

Keywords: Heraka, Sacralisation, Tingkupeu, Kelumki.

7.1 Introduction

The Heraka is a socio-religious reform movement of the Zeliangrong Naga. Heraka literally means 'pure' in Zeme, which is not mixing with other evil things.

The word *Hera*-means God and *Ka*- means fence. It means God fence against the evil forces and kept his people inside his fencing (Newme 1991: 1). Those who are inside the fencing, they are called Herakame, which means the pure people. It believes in the supreme God *Tingkao Ragwang* (God of Heaven), the creator of everything, and does not believe in the worship of spirits or smaller deities. This reformed religion aims in the abolition of the obscurantist customs and superstitious practices. The article is an attempt to examine the sacralisation of space in the religious reform movement of the Heraka. The study primarily involved an ethnographic method. The study is also based on participant observation method by interacting with the religious leaders, leaders of the Heraka organisation, and some important Heraka populations, attending in their religious functions, house of prayer for enriching in the study of their socio-religious realities. The study is engaged in qualitative method; therefore no attempt has been made to analyze them in quantitative terms.

7.2 The Beginning: Haipou Jadonang

The Zeliangrong consists of the three kindred tribes of Zeme, Liangmei and Rongmei. The Zeliangrong Naga is one of the various groups of Mongoloid race with distinct culture, laws and customs. According to some linguists and anthropologists, the Zeliangrong Naga belonged to the Tibeto-Burman family and sub-families of the Tibeto-Chinese race. The Zeliangrong Naga inhabit the compact and contiguous geographical area which has been fragmented into the states of Assam, Manipur and Nagaland in India. In Assam, the North-east of North Cachar Hills district is the ancestral Zeliangrong homeland. They are also settled in Cachar valley of Assam. In Manipur, Tamenglong district is the ancestral homeland of the Zeliangrong people. They are also settled in western Sadar Hills, Tadubi sub-division of Senapati district, Loktak Project area of Churachandpur district, certain villages in Bishnupur district, and about seventy villages in Imphal valley of Manipur. In Nagaland, Peren district is the ancestral homeland of the Zeliangrong people. They are also settled in Dimapur district and Kohima district.

Jadonang was the founder of the Heraka movement. He was born in 1905 in the village of Kambiron, in the present state of Manipur. The exact date of his birth was not known. At the time of his death in 1931, he was around 26 years, counting being done on the basis of the jhum cycle of his village and that he was 10 years older than his disciple and successor, the legendary Rani Gaidinliu who was born in 1915. Jadonang was born in a poor peasant family. His father was Thuidai of Malangmei clan, and mother Chunlungliu from the Dangmei clan. The British projected him as the freedom fighter of the Naga, while the Christian projected Jadonang as the leader of heathen movement, as they found him a stumbling block in the spread of Christianity.

Jadonang was conscious of the conservative, orthodox and superstitious religious beliefs of the Zeliangrong people. The Zeliangrong people believed in the idea of a Supreme God who was the creator of the universe, dispenser of good and prosperity and who lived in heaven. There are three categories of deities worshipped in the Zeliangrong traditional religion: the universal God and the gods of the lower realm, the deities presiding over the villages and the ancestors of the family. They also worshipped the elements of nature, the fire, the wind and the earth, and offer sacrifices to these deities. The Zeliangrong people became too superstitious that the worship of *Tingwang* or *Tingkao Ragwang* (Supreme God) was submerged in the sacrifices. They observed gennas and religious taboos which had replaced the religious prayer. Religion became a mere performance of sacrifices and propitiation of spirits and observance of gennas ranging from the birth of a child, a piglet, puppy, or hatching of chicken, a kite and crow sitting over the roof, a dog climbing the low roofs, burning of house, earthquake, still born child, death of women at delivery, death by drowning, and hanging. Despite such social and religious chaos, the rich and those who could afford still performed the religious sacrifices. Jadonang believed that *Tingwang* or *Tingkao Ragwang* (Supreme God) chose him to reform the superstitious and genna ridden religion of the Zeliangrong people.

Broadly, three stages of Jadonang's personality development are attributed: spiritual, social and political. The spiritual inclination was inborn which was evident in his initial life. From childhood, the behaviour of Jadonang was very uncommon, unusual and extraordinary. For an instance, Jadonang used to have religious trances sometimes for five days, and even for ten days. The frequent trances were considered as long sleeps. Then he began his religious activities. He constructed temples, and showed the people how to worship the *Tingwang* (Zeliang 2005:1-6). He also constructed *Tarang Kai*, a ceremonial house dedicated to God. He went on pilgrimage to holy places specially Bhuvon Hills in Assam. He also constructed temples known as *Kao Kai* to worship God.

In the social life, Jadonang was extremely afflicted to see the miserable and pitiable social hardship of the Zeliangrong people. That was the reason for his religious reform. The Christian missionaries under the British administration propagated Christianity in many parts of the world. During the time of Jadonang, some Zeliangrong people pledge to accept Christianity just to appease the British government for its favour. Jadonang was contemplative about Zeliangrong culture, religious beliefs, and tradition would be in grave danger if such religious conversion was to go unabated.

This is because the converts abandon their tradition and culture. Therefore, Jadonang established the tradition of temple construction in order to rouse and organize the people to avoid such conversion and made every temple a good centre of social organization. Different traditions of the village were made one common to all and the people were brought to stand in one platform of religion. Thus through social renaissance, Jadonang put forward a new dimension and cultural vitality to the society which was suffering from rigidity, narrowness and compression for a long time.

In the formation of Jadonang's political leadership, his visit to Messopotamia during the First World War was remarkable. It was perhaps from this adventure that Jadonang was inspired by the outside atmosphere to struggle for religious, social and political freedom. For an instance, the Naga Club was formed on the basis of the idea maintained by those soldiers who went to foreign land. Later on the idea of Naga National Council was established. After organizing the society and securing public support for his leadership, he attached himself constantly in making a plan to put a challenge before the colonial British. On the one hand, Jadonang tried to save his own society, culture and religion from the grip of foreign Christian missionary through constant spiritual thinking and social reform.

On the other, his concentration was fully confined to the attainment of political freedom which he called '*Makam Gwangdi*' or Naga Raj as Crosgrave, the then chief secretary of Assam called it (Kamei 2003: 145). The term 'Naga Raj' is referred to the Jadonang movement. This is perhaps, because the Zeliangrong belong to the Naga tribes, and as such the term 'Naga Raj' has been used by some writers. The word '*Makam Gwangdi*', means 'the kingdom of the Nagas'. He further even reiterated the slogan '*Makameirui Gwangtupuni*', which may be translated as 'the Nagas would rule one day'. It may be mentioned here that the word '*Makamei*' also suggest that Jadonang was referring to the Zeliangrong people in particular. Jadonang was arrested on 19th February, 1931 at Lakhipur by the British, on his return journey of the last pilgrimage to Bhuvon Hills, for declaring Naga Raj. Later he was executed by the British in the same year. Though he was the harbinger of a new reformed religion, he did not live to see the culmination of the reformed religion.

The reforms introduced by Jadonang in the traditional Zeliangrong religion were a synthesis of Christian monotheism and Hindu idolatry and temple culture. It was a revitalized and simplified form of worship. His philosophical reform was the worship of *Tingkao Ragwang* or Supreme God through prayer. He rediscovered the holy cave of Lord Bishnu in the Bhubon hills. He started the construction of temples where people worship God. The temple was called *kao kai*, and inside there was a shrine and pulpits and aisles of bamboo. The first temple was constructed in Kekru, then in Kambiron, Nungkao of Manipur and Binnakandi in Assam.

The abolition of enumerable taboos and gennas had purified the Zeliangrong traditional religion. He composed religious hymns to be sung during worship, and gave instructions for the composition of dances to be performed by the worshippers. These hymns, songs and dances were revealed by God. He constructed two temples, the first one like a church pulpit and articles in the central hall where he used to pray and preach. In the second temple he introduced a shrine with the temple, where he kept the clay idols of God Bishnu and his wife and a mithun. The first temple depicts the outward influence of Christianity in the church like temple for religious congregation and services, while the second temple depicts the influence of Hinduism.

7.3 Rani Gaidinliu: The Second Phase

After the martyrdom of Jadonang, Gaidinliu, the charismatic spiritual and political successor of Jadonang continued the legacy of Jadonang. The arrest, trial and execution of Jadonang stunned his followers with fear and anxiety. Jadonang's role was the preparatory phase and real action came during the phase of Gaidinliu, his brilliant and determined disciple. The visit to Kambiron (Manipur) around 1926-1927 was the beginning of a master and disciple relationship between Jadonang and Gaidinliu. She realized the spiritual power of Jadonang, and wanted to learn from him, and she became devoted to his master. Though, their association was only for about four years, Gaidinliu could follow the language and teaching of Jadonang. For an instance, Jadonang composed many songs and poems; Gaidinliu could learn them and also added her own in the later years. During the last pilgrimage to Bhubon cave, Lord Bishnu revealed to Jadonang and Gaidinliu, a new reformed religion, which is called Heraka.

Gaidinliu reorganized Heraka followers and challenged the British inspite of the innumerable odds and struggle. By her spiritual strength and skill in organization, she could convince the people and move forward to accomplice the mission of Jadonang. Several attempts to arrest her were failed for sometimes as the troops were allured that she appeared in different places simultaneously but actually she moved from place to place under the protection of her disciples. Troops were sent to Zeliangrong territory from all directions. At last, she was arrested in early 1933, tried and sentenced her to life imprisonment for starting such a heinous cult and carrying out insurrection against the British government, although it did not wholly succeed in suppressing the ideology on which it was built. She was released from Tura Jail when India became independent in 1947. Whatever might have been her teaching, Gaidinliu's heroism had moved veteran Indian nationalist Pandit Jawaharlal Nehru, who gave her the title of Rani in 1937. Her political programme was the translation of Jadonang's idea of the establishment of a Naga Raj. She continued to work for the welfare of the Zeliangrong people till her last breath. The important legacy of Gaidinliu was the reforming of the traditional religion which is known as Heraka.

The objective of the Zeliangrong movement under the leadership of Rani Gaidinliu was defense of her indigenous religion; revival of the political fight for the integration of the Zeliangrong into a common homeland. The Naga national workers in their political campaign took up Christian proselytism as a main function. Many of the Zeliangrong people who followed the indigenous religion were condemned or coerced to convert to Christianity. They formed into the Zeliangrong army and had a force, which in the course of six years of active existence (1960-66) reached a strength of 400 combatants and 1000 non-combatant civil followers. They came to be known as Rani Party, parallel to the Federal Government of Nagaland. The Zeliangrong government of Rani Gaidinliu was engaged in the twin objectives of preservation and promotion of the Heraka cult against the Christian preaching and establishment of a Zeliangrong administrative unit, either as a district or union territory. They were not opposed to the Naga independence movement as such, but their clash was more on the religious issue. The Rani party had not only created an army but also a civil government where Rani Gaidinliu was the patron saint chief of the Heraka religion and chief of the government.

Besides, being a freedom fighter and prophetess, Gaidinliu was a socio-cultural reformer. She was against the Western influences on the culture of the people. It was on this ground that she opposed Christianity as foreign religion that destroyed the traditional religion. She always stood for the preservation of indigenous culture and identity. Basically, it was for this reason that she always appeared on traditional attire. In one instance, while she was in Lucknow in her traditional attire, a Zeliangrong youth saw her and called her *apui* (mother). Then Gaidinliu told him that 'today you could recognize me because of my traditional attire, otherwise you would not recognize me, so wear our traditional dresses so that we could recognize one another'. To the present day, the Heraka instructed men to put on bronze earrings, and women to have short hair in the front side, so that they could recognize their fellow Heraka.

7.4 Sacralisation of Space-Kelumki (House of Worship)

Kelumki means house of worship. The *Kelumki* is separate from the rest of the village in the way it is conceived, as physically treated and respected. Every Heraka village must have one *Kelumki* and it is the most sacred ground for the Heraka. It is usually built on the highest point of a village. It is done in the same model of the first temple built by Haipou Jadonang and Rani Gaidinliu invoking as a chain of memory for legitimation. Every *Kelumki* must have a *Naimik Kekelum Be Bam*, place of sunrise prayer outside facing east. East is vital for two factors: it signifies the direction of Bhubon cave, as well as that of the sunrise, which is great importance during *Jalua*, full moon ritual. The sunrise prayer place must be elevated and is reached by the three steps symbolizing life-death-life. There is a veranda before entering the *Kelumki*. The main entrance is called *Muidi*; once inside there is another side entrance called *Muicheiki*, small door on the right, as it is strictly forbidden to place it on the left. In using the *Muicheiki*, one must re-enter using the same door.

Before entering the compound of the *Kelumki*, shoes must be removed. Inside there are two sections: the right for the male and the left for female. On the left there is a bench facing the congregation for the girls specially chosen to serve the Heraka community and on the right is reserved for the secretary, chairman, president, and priest of the village. Podiums are also present on both sides for those wanting to say a few words. On the podium is written '*Tingwang Hingde*' (God's law) and below that '*Sam Yi Besa Bam*' (Speak the truthfulness). At the front of the *Kelumki* is the altar, centrally placed and elevated. One must climb the three steps to it, proceeding the right leg, then left, and right.

The altar is on two levels. On the first level, only fruits, vegetables and flowers can be placed while the second level is for coins and paper money. This offering is accompanied by a prayer (personal and private). Once that is finished, one must turn anticlockwise and return to one's place. While one waits to reach the altar, a space must be kept between the right and the left for *Tingwang* to come and go in between. When one looks directly above from the altar, there is a small opening into a short wooden tower.

The way of worship in the Heraka religion is very simple. It is a devotional worship. In the Heraka religion prayer may be performed individually or collectively at any time and day, whenever require. An individual may offer prayer every day in the morning and in the evening usually before sleep. The devotees compulsorily offer sunrise prayer on every Full Moon day and the Heraka New Year Day at *Kelumki*. An individual or collectively offers prayer at the time of ailments and birth or death of a person. They pray at *Kelumki* on the Full Moon Day and the Heraka New Year Day. In the early morning, they gather at *Kelumki* and recite hymns and offer collective Sunrise Prayer. The Priest offers prayer to *Tingwang*. Then, they enter *Kelumki* in singing devotional song. Individual member also offers prayer at the holy altar. They sing a rejoicing song in praise of *Tingwang*. On the Heraka New Year Day, they sing a devotional song followed by incantation of Heraka hymn and drink *Telau dui* (holy water). The *Telau dui* is used as sacrament to cure or prevent sickness and to cleanse sins on the full moon day. It is believed that *Telau dui* is the holiest water which is available only in Bhubon Cave. Therefore, Rani Gaidinliu used this *Telau dui* for healing the sufferings of her followers during her freedom struggle against the British. The Heraka people use to collect *Telau dui* from Bhubon Cave every year. It is used on the inaugural day of a new house of worship and sprinkling of *Telau dui* upon the person indicates he is converted into Heraka religion.

7.5 Institutionalisation of Priesthood (Tingkupeu)

In most of the Naga villages there are priests who perform religious duties and represent the whole community in worship. Their main function is to offer sacrifices at all festivals and other important occasions such as births, deaths, marriages, and making the deity happy. The priest also communicates with the dead person's soul and delivers messages to the family. This communication is possible only when the priest goes into a trance. It is also the duty of the priest to schedule the day for general worship by counting the days, months and seasons (Shikhu 2007: 15).

In Heraka the priest plays a very important role in the religious life of the community. The priest is known as *Tingkupeu*, which means the one who invoke or call on *Tingwang* (God). Priesthood in Heraka is not hereditary but depends on the quality of a person. There are criterions to become *Tingkupeu* such as humility, obedience, righteousness, love, holy, kindness, and faithful. The *Tingkupeu* performs rituals and prayers on behalf of the village in *hejuadekung* (the sacred altar), as the intermediary between God and men.

Besides performing rituals and prayers in *kelumki* (house of worship), the *Tingkupeu* also teaches about the Heraka religion to the people. Recently, the followers of Heraka have instituted a preacher who is known as *Hingde Paume*, whose task is to go to places and preach the Heraka religion. Like the Christian missionaries, the *Hingde Paume* goes to villages to preach and teach the people, the law of Heraka religion. However, the *Hingde Paume* does not go to Christian villages, but only to *paupaise* or Heraka villages. Therefore, the concern of the Heraka is to protect and preserve their religion. *Paupaise* is the primordial religion of the Zeliangrong people. There are still few *paupaise* households in villages like Hezailoa, Kipeiloa, Tungje Punggo (Hereimidalo), and Hungrum of Tamenlong district. The Heraka frequently visits these places to bring them into the Heraka fold.

However, there are some instances where Christians are re-converted to Heraka. Those re-converts are mostly people who became Christians to get benefit from conversion. When their expectations of benefits are not materialized, they go back to Heraka. Those re-coverts to Heraka are awarded with gifts and prizes from the Heraka community. Christians claim that the institution of *Hingde Paume* (Heraka preacher) is nothing, but an imitation of the Christian missionaries. Christians also argued that whenever Christians have a programme of evangelism in villages, the Heraka preachers used to come in advance in the villages and dissuade the people not to attend the Christian meeting. The Heraka preachers even suggested the people not to listen, in case they attend the Christian meeting. Thus, the *Hinde Paume*, apart from teaching and preaching Heraka religion, also engage in dissuading the people from attending Christian meetings.

The *Tingkupeu* offers prayer in occasion like house dedication, marriages, and while attending the sick. Normally the *Tingkupeu* offers prayer for the sick, visiting their homes. If the need arise, then group of priests would offer prayer for the sick in *paiki* (supreme body in the village polity). This is because the *kelumki* (house of worship) is never open for any emergency cases. The *kelumki* is open during full moon day (*jalua*), and during festivals. The *Tingkupeu* also offers prayers while naming a child. If a child is born during daytime, the child is brought to the sun for prayer. If the child is born after sunset or at night, then the child is brought to the sun the next day. The significance is that the child is a gift from God. It is an act of honouring God that he has given a child. However, this prayer can be offered even by an elderly woman, in absence of the *Tingkupeu*. After five days, the *Tingkupeu* prays for the life of the child; to be successful and become a useful person in society. If the child is dead before five days of its birth, even then funeral is performed the same way it is performed for an adult.

On the full moon day (*jalua*), the *Tingkupeu* offers prayer when the sun rises in the morning in the *kelumki*. The people along with the *Tingkupeu* offer prayer to the sun, invoking *Tingwang* so that they could be like the sun, giving light into this world. Before the full moon, the priest and the people have to keep themselves clean one day ahead. The *Tingkupeu* is not allowed to kill any living creatures or pluck any plants. Besides, he should not share bed with his wife and abstain from worldly things. He should fast and remain clean. On the full moon day, he should remain clean. After the prayers and rituals, the *Tingkupeu* can break his fast.

The *Tingkupeu* is the permanent priest of the village. He is assisted by one junior priest. Whenever the *Tingkupeu* is ill, his assistant performs rituals and prayers in the village. If the *Tingkupeu* fails to perform his duty well, the villagers can remove him from priesthood. He can be also removed if he is morally corrupted. If he remains good and faithful, he can remain as priest for his lifelong. Besides, the *Tingkupeu* can step down from priesthood if he is too old, and finds it difficult to carry out the rituals, prayers and ceremonies. Every Heraka village has one *Tingkupeu* with one junior assistant. Every matters relating to the *Tingkupeu*, like selection or removing him from priesthood are decided by the villagers in their respective *paiki*. The *paiki* decides and governs the institution of priesthood.

Nevertheless, the *Tingkupeu* is the head of the village. He plays a leading role in all religious activities of the village. He is regarded as a commune with God. So, he is respected by all as Divinity. The priest of Heraka is selected to the oldest male from amongst the villagers who has courage enough and experienced in worshipping to control his villagers and having fanatic childlike minded person. When the *Tingkupeu* is once selected then he has to govern the role of priest till his long life time. The *Tingkupeu* and the holy altar stone must be properly maintained and should be considered it as the Chief and the dependency of the village respectively. *Tingkupeu* has to perform every prayer of the village such as to perform prayer for good health and rich harvest of the village both in individual or collective affairs. He will broadcast the day of prayer on the eve of every prayer and perform prayer on every important festival.

7.6 Conclusion

The Heraka is a socio-religious reform movement of the Zeliangrongs derived from their traditional religion. The reform aims in the abolition of the obscurantist customs and superstitious practices. There are scholars who argued that the Heraka is a debased form of Hinduism, while others tries to maintain the Heraka distinct religious identity. The most important aspect of the Heraka reform, among others, is the institutionalisation of priesthood and the sacralisation of a place of worship. This is because, for a long time, the Zeliangrong people have no system nor particular time and place for regular worship. They have no congregational service and no recitation of any kind since there is no proper religious literature. There is no confession, and they worship the Supreme Being only in times of distress and sickness. Certain places, however, are given due reverence. Sometimes they pay reverence to certain trees which are believed to be the abode of a supernatural power. Besides, the role of priests is minimal and less significant. The religious reform, therefore, besides others, brought certain institutions into a proper system.

References

Kamei, Gangmumei. 2004. *The History of Zeliangrong Nagas: From Makhel to Rani Gaidinliu.* Guwahati: Spectrum Publications.

Longkumer, Arkotong. 2008. *Where Do I Belong?: Evolving Reform and Identity Amongst the Zeme Heraka of North Cachar Hills, Assam, India.* (Unpublished Ph.D. Thesis, Religious Studies Department, School of Divinity, University of Edinburgh, May).

Newme, Pautanzan. 2004. *The Basic Facts of Heraka Religion.* Zeliangrong Heraka Association, North-east India.

Newme, Pautanzan. 2002. *Silver Jubilee Souvenier.* Zeliangrong Heraka Association: Assam.

Newme, Ramkuiwangbe. 1991. *Tingwang Hingde.* Guwahati: Regional Zeliangrong Heraka Association, Assam.

Shikhu, Inato Yekheto. 2007. *A Re-Discovery and Re-Building of Naga Cultural Values: An Analytical approach with special reference to Maori as a colonized and minority group of people in New Zealand.* New Delhi: Regency Publications.

Zeliang, N. C. 2005. *Zeliangrong Heraka Movement and Socio-Cultural Awakening in Naga Society.* Zeliangrong Heraka Association: Manipur, Assam, Nagaland.

Zeliang, Thunbi. 2005. *Haipou Jadonang.* Gauhati: Heritage Foundation.

Papers Published by Various Organisations on Rani Gaidinliu

2007. *Rani Gaidinliu* By Nagaland: Zeliangrong Mipui Organization.

Religious Conception of Rani Maa Gaidinliu by Zeliangrong Heraka Association.

2000. *Silver Jubilee Souvenir.* Kohima: Zeliangrong Heraka Association.

2009. *Zeliangrong Customary Laws.* Haflong: Zeliangrong Heraka Association.

2005. *Zeliangrong Heraka Phung Kesubo.* Kohima: Zeliangrong Heraka Association.

About the Author

Soihiamlung Dangmei is Assistant Professor, in the Department of Political Science, Indira Gandhi National Tribal University, RC- Manipur. He has published few articles in reputed journals, and also has presented some papers at national and international conferences.

E-mail: soihiam@yahoo.com

Chapter 8

Festivals and Rituals of the Maram Naga Tribe

☆ *Peter Ki*

ABSTRACT

The study is an exploration of the festivals and the rituals of the Maram Naga tribe of Manipur. Though numerically small, the tribe has a deep reservoir of traditional and cultural practices. Fortunately, few people continued to adhere to these old practices, even though the majority have converted to Christianity. Of late, attempts, with a reasonable degree of success, are being made by the people through its frontal organisations to keep some aspects of tradition and culture alive.

Keywords: *Maram, Naga, Festivals, Rituals, Mangkang, Punghi, Kanghi.*

8.1 Introduction

The Maram Naga tribe is one of the myriad tribes of North-east India.[1] They inhabit in Senapati district of Manipur. They belong to the Naga ethnic group. Though numerically small, the Maram has a rich cultural heritage.[2] They are spread over thirty-nine villages.[3] They speak a language/ dialect known as Maram.[4] The Maram is surrounded by other Naga tribes like the Mao Naga on the North, the Poumai Naga on the East, the Thangal Naga on the South, and the Zeliangrong Naga on the West and South West. With the advent of Christianity and modernity, traditional and cultural practices are on the declined. However, efforts are being made by the tribe's frontal organisations to preserve its traditional and cultural heritage.[5] Majority of the people have embraced Christianity, but a small number of people still professed to the traditional practices and beliefs, which many considered to be a form of animism.[6]

8.2.i Traditional Rituals and Practices

It is through the enduring influence of this small number of Maram people who are still committed to animism that traditional practices and rituals are strictly observed, especially in Maram Khullen village. However, only few observed the traditional practices and rituals, the majority of the people have converted to Christianity. Nevertheless, this village continued to play the role of preserver of the tribe's culture, tradition, social norms and its ethos in the basic form. Its residents zealously guard many of the customs and traditions even though most of them have migrated to different to places and villages.

The Lunar calendar is followed for the observance of the traditional practices and rituals. For many Maram, whose lives revolve around a tiring and still largely on primitive mode of agricultural cultivation, celebrations and rituals are kind or another throughout the year that serves as a much-needed respite.

8.2.ii Major Festivals of the Maram Naga Tribe

There are two major festivals of the tribe: *Punghi* and *Kanghi*.[7] These are celebrated after completion of rice transplantation and harvest respectively.[8]

Punghi – Post-Plantation Festival

Punghi is celebrated in July.[9] This festival spanning seven days of celebration marks the end of the transplantation period, generally considered as the most difficult phase of rice cultivation. Therefore, the festival is a celebration after days of hard work. Rituals are observed for good harvest. During the celebration, there is a lot of merry-making, involving generous consumption of meat and rice beer.[10]

The festival is also known as *Punghi kiibah*, which literally means bright *Punghi*. It is an indication that with this festival the Monsoon season ends and bright sunny days begin.

Kanghi – Post-Harvest Festival

Kanghi the biggest festival of the Maram Naga tribe is celebrated in December (*Kanghikii*) for seven days with merriment. A special feature of this festival is wrestling, in which a few men even today engage in naked wrestling. On the fourth day of the festival, wrestling takes place in front of the house of the king of the Khel.[11] On the fifth day, it is performed in front of the house of the King of Maram Khullen. Naked wrestling still takes place with a belief that evil spirits will not leave the people in peace without it. During this festival, the people gather in groups to sing melodious folk songs and listen to folk tales.

It is also referred to as *Kanghi Kajing*, in which *Kajing* literally means cloudiness or unclear skies just before a rain or a storm. The festival marks the end of bright sunny days and heralds the arrival of cloudy or gloomy days when the process of rice cultivation will start all over again.

Immediately after *Kanghi* festival *Kanghi Maru* a day of ritual cleansing is observed. On this day, the men folk perform ritual cleansing by taking bath early in morning from water ponds located near the village. After the cleansing, boys go for hunting and the meat from the spoils is shared as per the traditional practice.

8.2.iii Other Festivals and Rituals

Apart from the two major festivals discussed above, there are other minor festivals and rituals. Of particular interest is a festival dedicated for the women folk of the tribe.

Mangkang – A Festival for the Womenfolk

In spring season, usually corresponding with the month of April[12], the Maram women folk are treated to a festival known as *Mangkang*.[13] The festival is unique for the reason that it is exclusively celebrated for the womenfolk.

The feast day of *Mangkang* is preceded by elaborate preparations lasting four days. On the first day (called *Kaliimra*), the men folk go to the jungle to collect the bark of a particular tree (*taking-ting*) to use for fishing; on the second day (called *Matam*), the tree bark collected the previous day is pounded into powder; on the third day (called *Mala Kasa*), the men folk go to the river for fishing. And on the fourth day (called *Mala*), rich families treat the children of poor neighbours to a meal of fish, chicken, snail seasoned with ginger leaves.

The day of *Mangkang* is like welcoming the New Year. All kitchen utensils are washed, and the fire place made anew. Food items such as snails, prawns, crabs, fish, and chicken seasoned or flavoured with fresh ginger leaves are prepared on this day too. The girls then gathered at a common dormitory (called *raliiki*) and relish the delicious food each has brought from their own homes. They also go to bathe in groups to prepare themselves for the evening festivities.

Finally, when evening arrives the young unmarried girls (along with the girls who married during the year), wearing traditional attires such as necklaces, bangles and earrings, gathers at a traditional place called *Psiihapung* to perform the traditional dance called *psiiha*. The girls engaged themselves in dancing competition to determine who the best among them is. The festival is, in a sense, a platform for the women folk to showcase their wealth, beauty, and agility. This festival is equally important for the young unmarried boys too. Because it is an opportunity to openly interact with girls without inhibition - being socially sanctioned - helps them choose their life partners on such occasion.

Mangkang is followed by *Tingpui Maru*, a day of ritual cleansing, similar to *Kanghi Maru* described earlier. On this day too, a hunting expedition is undertaken.

Atiim Matai

Even the cow, reared for meat is the centre of a day celebration called *Atiim Matai*, celebrated in October. During this celebration the cows are given names. On the very day, in the morning mostly children below ten years, under the guidance of the cowherds are engaged in social work and clearing the stones from the cow paths. In the afternoon, when the task are done the children are treated to a meal with rice and chicken along with rice beer.

Maliim

Maliim is a ceremony of ritual cleansing observed for a day usually in December. This ritual is performed to ensure that no injury or ill befall on the villagers. In the past, a successful observance of *Maliim* was taken as a good omen, and that if a war breaks out that year the people would be triumphant over her enemies in that war.

Only the men folk are involved in this ritual. Early in the morning, the men folk gathered at the village resting place (*ranii*), and wait for the sun to rise. If the sun rises without any cloud blocking its path, it is considered as a good omen, and they proceed to the jungle to hunt for birds and small animals, and undergo the process of ritual cleansing.[14] Once they return to the village, the men folk retreat to their respective traditional boys' dormitories (*rahangki*) and feast on rice and chicken cooked in earthen pots.

N'pamrah

Infants performed two ceremonies in a year. In this ceremony, wine is offered to the elders by every infant's household. In return, the elders bestow them blessings.

Riikak

In November, a ritual called *Riikak* is performed in memory of all the departed souls.

8.3 Conclusion

Today, the indigenous traditional practices and rituals of Maram though on the verge of extinction are still observed by few people who have not converted to Christianity. With the increased conversion to Christianity and adoption to modern way of life many have become less interested in their traditional way of life. There are attempts, with reasonable degrees of success being made by the tribe's frontal organisations to retain at least some significant traditions and customs alive to uplift the distinctive identity.[15]

Endnotes

1. The study is based on primary sources, largely on the author's personal interactions with elders and knowledgeable people of Maram society. The study primarily focuses on Maram Khullen village also known as *Maramai Namdi*, the oldest village where traditional practices and rituals are still observed strictly by few non-Christians. There are very few literatures on Maram tribes. The author would like to acknowledge Mrs. Hoidina Lucy for proof reading the draft.

2. According to UNESCO database on endangered languages, the number of speakers of Maram language is 37,000. However, the official data is not available as the census results for 2001 and 2011 for Mao Maram, Paomata and Purul sub-divisions of Senapati district of Manipur were cancelled.

3. This is based on a Souvenir titled '*Lui-Ngai-Ni 2010: The Naga Seed Sowing Festival*', published to commemorate the celebration of Lui-Ngai-Ni 2010 on 15th February 2010 at Senapati (Tahamzam) with the theme "Our Resources – Our Future".

4. Under the UNESCO's classification of 'degree of endangerment' of languages, Maram language has been categorised as 'vulnerable'. It means that "most children speak the language, but it may be restricted to certain domains (e.g., home)". So, it is not that bad if one were to look at it in terms of the spectrum of degree of endangerment which ranges from 'vulnerable' to 'extinct'. If a language is to be categorized as 'extinct', it means that "there are no speakers left". The Maram language is not in immediate danger of extinction, but if the current level of negligence continues the alarm bell might ring soon.

5. Some of the prominent organizations engaged in this task are the Maram Union (the apex body of the Maram Naga Tribe), Maram Women Union, and the Maram Students Union.

6. Like animism, the Maram Naga believed that personalized supernatural beings (or souls) inhabited certain objects and governed their existence. For example, there is a big stone on the outskirt of Maram Khullen village, known as *Apou M'raba* (Grandfather *M'raba*), and is revered as a good soul.

7. The information of these two festivals; *Punghi* and *Kanghi* was provided by Dishung-pui Poina of Maram Khullen village.

8. However even after the celebration of this festival, those who could not complete the rice transplantation can continue to do so.

9. The corresponding month in the lunar calendar of the tribe is *Punghikii.*

10. The tradition of merry making and meat consumption is carried on by the people irrespective of their place of residence, although consumption of rice beer is limited to few villages.

11. Maram Khullen is divided into three regions namely; *Kagamna (Makha Sagai), Lamkana (Khullakpa Sagai), and Magai-Bungnamai (Mathak Sagai)*. There is a chief for each Khel. However, the chief of Lamkana (Khullakpa Sagai) is considered as the king of Maram Khullen village and also for the entire Maram tribe.

12. According to Pungdi Celestine, the president of the Maram Students' Union (MKS), *Mangkang* usually takes place on the fourth day of the month of Tingpuikii (the fourth month of the traditional lunar calendar). In case of leap year it falls in March, and in normal years it's in April.

13. According to Mrs. Raina Phuba (a social worker) the festival owes its origin to the constant state of war-making in the past. Since the able-bodied men folk were required to defend the village, the girls were engaged as cowherds, which was quite a tiring task. In recognition of their hard work a day of feasting was thus set aside for them.

14. If the cloud blocks the path of the rising sun, it is considered inauspicious and the ritual is put off for another day. This ritual is taken very seriously. *Rang-pui Hinga* of Maram Khullen village recounted the tale of a particular year in which

the ritual was successfully observed only after seven attempts. This became a big problem for the people since the ritual is preceded by elaborate preparations that include brewing rice beer which took time to mature.

15. For instance, the *Maram Khullen Circle Women Association* (MKCWA) organised an annual festival of *Mangkang*, which mixes traditional dance *psiiha* with other activities such as Miss. Mangkang Contest and singing competition.

About the Author

Mr. Peter Ki completed M.Phil from Jawaharlal Nehru University (JNU), and has worked as Research Assistant at National Security Research Foundation, New Delhi. He has served as Editor-cum-Communication Officer at Economic Research Foundation, New Delhi and also as Assistant Coordinator at Euro-Burma Office, New Delhi besides briefly working as a Fellow Coordinator, Programmes at India Development Foundation, Gurgaon. Currently, he is Public Relations Officer at Nagaland University.

E-mail: pkmaram@gmail.com

Section – III

Rural Development and the Autonomous District Council for Hill Areas

Chapter 9

Rural Development in North-East India: With Special Reference to MGNREGA in Tamenglong District of Manipur

☆ *Pr. Dimchuiliu*

ABSTRACT

This presents chapter focus on the study of Tamenglong district of Manipur, on the nature of implementation and the variation of development. The chapter identifies how insurgency played a major issue with regard to conflict, impacted subsequent government policies and long-run economic development. The study also reveals that rural development in Tamenglong district was severely slowed down and affected due to rural backwardness and insurgency and counter insurgency groups. It also indicates that various rural development initiatives were introduced and implemented but its impact on rural economy is minimal resulting in insufficiency of village economy. The chapter focuses mainly on the impact of village's groupings and counter inters insurgency strategy and the rural development programmes initiated and its impact on rural district of Tamenglong NREGS.

Keywords: *Rural development, Backwardness, Tamenglong district, MGNREGA.*

9.1 Introduction

Development is defined as 'a function of capital and technology; development goals were cognized in terms of urban middle class aspiration. Development crystallized; it dawned on a few of the experts and bureaucrats as well as the people at large that to be authentic, development ought to be participatory'(Oommen 2004: 33). Development by definition is the act or process of developing or growth within the framework of available resources (Seers 1972). Sachs defines development as a weapon in the competition between the political systems (Sachs 1994:1-2). According to Weinder development is a state of mind, a tendency, a direction, rather than a fixed goal. It is the rate of change in a particular direction (Mishra and Sharma 1983:46). Development means making a better life and is a desirable goal of modernity with all modern advances in science and technology, in democracy and social organisation, in rationalized ethics and values, fuse into the single humanitarian project of deliberately and cooperatively producing a far better world for all. Development is optimistic and utopian. Development means stating change at the bottom rather than the top.

It is not only defining 'development' which is contested, the way that development, regardless of definition is measured is also problematic. Development is a specific process going on in most formerly underdeveloped countries. According to Bottomore there are 'only two (related) social processes to which it seems possible to apply the term development with any accuracy; namely the growth of knowledge and the growth of human control over natural environment as shown by technological and economic efficiency' (1986:286). It is indeed these two processes which have figured most prominently in developmental accounts of human society.

Development as a term also gained much currency in the literature pertaining to social dynamics. It has been synonymously used with modernization. There are however some scholars, especially those with Marxist perspective who believe that those term should be abandoned and put forth various argument to support their contentions. The term 'development' has sometimes been applied to mean economic growth process and 'modernization' to mean various socio-cultural processes concomitant with them (Pandey 1985:79). The idea of development according to Ogborn (1999) is linked to concepts of modernity. 'Modernity' in its broadest sense means the condition of being modern, new or up-to-date, so 'the idea of 'modernity' situates people in time (Kumar 2008). However, more specifically, 'modernity' has been used as a term to describe particular forms of economy and society based on experiences of Western Europe (Willis 2005: 2).

The very concept of rural development, though, being debated frequently is hard to define. Even among the great thinkers, no consensus has been achieved in this regard. The term 'Rural Development' is a focal interest and widely acclaimed both in the developed and developing countries of the world. There is no universally accepted definition of rural development and the term is used in many ways and in vast divergent context as it has been understood in different ways as concept, a strategy, a discipline, a process, a slogan, propaganda, a philosophy as rural industrialisation to create rural transformation, a structural change in the socio –

economic framework and as mechanism. But rural development has always been defined with reference to the development of rural areas with certain specific objectives. The concept of Rural Development is viewed from different perspectives by different people as it has undergone a tremendous change. And the concept of rural development was born in the context of agriculture and remained so for a long time, co – terminus with agriculture development. Right from the launched of First Five Year plan in 1951, rural development has been receiving the attention of planners. The concept of rural development in the beginning of Planning in India was confined to the development of agriculture and its allied sectors only. Since 80 per cent of the rural population is dependent upon agriculture the planner believed that the growth of agricultural production would lead to the welfare of the rural masses. However, on the other hand, rural development also generally refers to the production and utilization of material resources or the enrichment of human resources.

The concept of rural development connotes overall development of rural areas with a view to improve the quality of life of the rural people. In this sense it is comprehensive and multidimensional concept and encompasses the development of agriculture and allied activities- village and cottage industries and craft, socio-economic infrastructure, community services, and facilities and above all, the human resources in rural areas. Therefore, there is a need to define rural development in a wider perspective as dynamic process of societal transformation from a traditional to a modern society.

Rural development is polemic. And it was gradually realised that the rural life in India is beset with numerous other human problems and that the conditions cannot be improved merely by launching a few segmented economic programmes. Thus, there is an urgent need to go into the root of the matter and consider rural development as essential, primary and basic for developing nations, which have, has a vast rural population to cater for. Therefore, the popular notion of rural development needs to be converted into reality and should reflect the development of rural masses. This realisation led the thinkers to look into the problem of rural development through a broad angle to the totality of rural life, encompassing the entire scenario of social, cultural and environmental conditions together with the problems connected with the basis amenities of rural life.

In the words of Robert Chamber, "Rural Development is a strategy to enable a specific group of people, poor rural women and men, to gain themselves and their children more of what they want and need. It involves helping the poorest among those who seek a livelihood in the rural areas to demand and control more of the benefits of rural development. The groups include small scale farmers, tenants and the landless" (Chamber1983:20).

9.2 Profile of Tamenglong District, Manipur

Tamenglong District is endowed with rich natural resources which are essential for high economic uses of land. It located in the north-eastern corner of India, in the Hills of Manipur which extends from 25° 7' N to 25° 27 North Latitude and 93° 10' E to 93° 42' E Longitude. It is bounded by the state of Nagaland on the North, Cachar

District of Assam on the east, Churachandpur and Senapati Districts on the South and East respectively. It is a mountainous clad with thick forests except tiny Khoupum valley of 3 sq. Mile. The major rivers in the district are the Barak, the Irang, the Makhu and the Apah. The district has four sub-divisions; namely, Nungba, Tamenglong, Tamei and Tousem with different dialects. Each sub-division is co-terminus with blocks on the basis of administration by the State Government and these sub-divisions comprises 242 villages in the district. The total geographical area of the district is 4,391 sq.km or 19.66 percent of the total area of the state of Manipur. The population according to 2011 census is 1, 40, 143. Tamenglong District rank ninth in the state population of Manipur as par the census of 2011. The literacy rate of the district is 70.40 in which male constitute 76.74 and female 63.72 respectively.

9.3 Mahatma Gandhi National Rural Employment Act

National Rural Employment Guarantee Act (NREGA) which came into force in 2006 was launched in 200 India's most backward areas (Guidelines for implementation of works on individual land under NREGA GOI) Over the first few years of operation of the Mahatma Gandhi National Rural Employment Guarantee Scheme (MGNREGS) has been a blessing as number of development projects were undertaken in the rural areas. The objective of the Act is to enhance livelihood security in rural areas by providing at least 100 days of guaranteed wage employment in the financial year to every household whose adult members volunteer to do unskilled manual work.

NREGA guaranteed employment for at least one person from each household for 100 days under the scheme. And it is promising that NREGA commits to pay a minimum wages while banning both contractors and machinery from taking part under the scheme though use of machine are encouraged in some situation to reduce the drudgery of work. The scheme was also expected to generate better income as majority of the people in this area depends mainly on the wages they earn through unskilled, manual labour. NREGA contains several provisions for insurance against accidents, disability and death while on duty, provision of medical aid, drinking water, crèche etc for the labour under this historic development scheme. However, the challenge is to sustain and improve the implementation in order to reduce the problem of poverty and unemployment.

NREGA is supported by an extraordinary set of guidelines issued by the Ministry of Rural Development, Government of India [MoRD 2005a]. The Act also seeks to put into place elaborate mechanisms for preventing leakages and corruption. And it is committed to enhance the productive capacity of the rural economy through creation of durable assets. NREGS is a development scheme to gain momentum of growth in the most backward regions of the country. The scheme emphasis on the following kinds of work as per Schedule I of the Act:

☆ Water conservation and water harvesting;

☆ Drought- and flood-proofing including afforestation and tree plantation;

☆ Irrigation canals, including micro and minor irrigation works; provision of irrigation facility, plantation, horticulture, land development to land owned by households belonging to the SC/ST, or to land of the beneficiaries under the Indira Awas Yojana/BPL families.

⭐ Renovation of traditional water bodies, including desilting of tanks;

⭐ Land development;

⭐ Flood control and protection works, including drainage in waterlogged areas;

⭐ Rural connectivity to provide all-weather access. The road construction of roads may include culverts where necessary, and within the village area may be taken up along the drains. Care should be taken not to take up roads included in the PMGSY (Pradhan Mantri Gram Sadak Yojana) network under NREGA. No cement concrete roads should be taken under NREGA. Priority should be given to roads that give access to SC/ST habitations;

⭐ Any other work that may be notified by the Central Government in consultation with the State Government.

It is also well known that the National Rural Employment Guarantee Act (NREGA) addressed many of the weaknesses of earlier programmes through several features in its design. NREGA introduced a rights-based framework;

1. It introduced a legal guarantee of work, as opposed to a government programme which could be withdrawn by a government at will;

2. Time bound action to fulfil guarantee of work within 15 days of demand for work;

3. Incentive structure for performance (central government funds 90 per cent of costs of generating employment);

4. Disincentive for non- performance (unemployment allowance to be paid within 15 days if work not provided within 15 days is a state government liability);

5. Demand- based resource availability; and

6. Accountability of public delivery system through social audits.

9.4 Problems of Rural Development in Tamenglong District

Tamenglong district of Manipur which is considered one of the most backward districts [as per the MGNREGA in which 200 most backward district were covered in the first phase (Guidelines for implementation of works on individual land under NREGA GOI)] is also faced with multiple serious problems of topographical location, illiteracy, local politics, corruption, and insurgency issue. As a result of these issues, investment on development scheme is poor and minimal in this region. Resources in this area are scarce and demand is greater than supply rate. The recipient of development schemes is mostly poor, ignorant and illiterate. The rural people do not have the power to challenge since they are illiterate and lack the political power and as a result of this they suffer mostly in the hands of insurgents and corruption. Even the well educated literate people lack the power to stand up for their rights and fight against this social evil, but, instead they succumb to it. So, in this case, Athui Gangmei said, 'if the leaders who know the roots of the entire socio-economic problem keep things to themselves rather than standing up for the cause of the weaker section, is there any scope for change and development in the so called backward society? There is much talk with regard to rural development but, where is the solution to uplift the rural folk? Is this all that the government have to say?'

Theoretically, there is much implication toward the upliftment of the rural areas especially like Tamenglong District. But practically where are those promises? With reference to the MGNREGS the beneficiaries were supposed to be given security even in the work place by providing shade, safe drinking water, crèche, notice board..., regular payment of wages, transparency in its implementation and decision making. Taking all these issues into consideration rural development in Tamenglong district is like a delusion. In spite of being identified as the most backward district in India, the state and central government turns their back. From the study conducted in Tamenglong region, particularly in the three sample villages of my study it is visible that there has been much manipulation done in the paper work. The poor and illiterate people have thumbed away their rights and even their traditional holdings like land in that regard (A case in Khongjaron village). There is much of exploitation and discrimination mushrooming in the small district. The concept talked about by Marx on the notion of 'Haves' and 'Have Not' is visible with the emergence of class in the previously classless society.

The so called Tamenglong District is backward and is also the hub for the insurgent groups. The public suffers at the hands of all the higher authority. Be it, village authority, church, politicians, bureaucrats, insurgents and even the landowners and money lenders. Thus, implementation of rural development is practically complicated in this district.

9.5 MGNREGS in Tamenglong District: Implementation

On February 2006, the National Rural Employment Guarantee Act (NREGA) came into force in 200 of India's most backward districts. Tamenglong district is among the districts covered when it was launched. The operation of the Mahatma Gandhi National Rural Employment Guarantee Scheme (MGNREGS) in the district has been a blessing as number of development projects were undertaken in the rural areas. And also it enhance livelihood security in rural for the people by providing at least 100 days of guaranteed wage employment in the financial year to every household whose adult members volunteer to do unskilled manual work.

Though list of permissible kind of works are laid down, it is also critical to guarantee implementation due to Geographical location and seasons. The feasible works in Tamenglong District includes water conservation, horticulture, renovation of traditional water bodies (de-silting of tanks), road construction and drainage as the pre-requisite and foundation for rural transformation.

The current milieu of MGNREGS in Tamenglong is the employment generation, which has come to occupy centre stage in development planning as well as implementation. It brings forth growth and structural change in the economy by expanding productive employment to the unskilled population as labour is the main asset for a majority of the poor and sustained them by reducing the level of poverty.

Apart from direct benefits of employment, the scheme also fulfils the function of unemployment insurance. There is, how-ever, debate on the productivity of the assets created. The categories of works that are undertaken under NREGS in this area are: water conservation and water harvesting; renovation water sources; flood control and protection works including making of drainage; and rural connectivity to provide all-weather access. All these activities improve rural infrastructure of the district. Apart from the direct benefits of employment and incomes, the scheme has many indirect benefits such as seasonal benefits, insurance function, impact on agricultural wages, and increase in women's employment, making rural poor as a political force and, organising the unorganised workers.

The main actors of the implementation and monitoring authorities in the district are: the district coordinator, the programme officer, the village council. People are cynical about the proper implementation of the scheme as there is no adequate capacity to implement without leakages and corruption. This is partly because of the lack of people's participation in decision making and due to the problem of insurgency in the district. The delivery system is meagre as social mobilization, right to information and involvement of the public is minimal.

Although there is much to talk about the benefits of the scheme, there is a negative side of it. Since, after the implementation of NREGS there is much complication in the village level as people craves for more financial and political power and leadership. The elected/selected leaders of the respective village were challenged by those who have more political power. Even the family relations become edgy and the bond of unity which was previously present started to take an ugly turn. The character of simple society or rural society has lost its simplicity as they see more money approaching to their doorsteps.

Rural development with reference to MGNREGA in Tamenglong district was found to be missing out in many important aspect of the scheme in the process of its implementation. The facilities which were supposed to be implemented and provided to the beneficiaries were totally missing which can be illustrated as under.

9.5.i Worksite Facilities

Worksite facilities such as crèche, drinking water, shades and first aid facilities which are supposed to be provided are totally missing. Likewise, a grievances redressal mechanism for ensuring a responsive implementation process is absent. The beneficiaries are not aware of the facilities that are to be provided to them under the scheme because most of them are ignorant of the Act and the benefits under the Act.

9.5.ii Disbursement of Wages

In spite of the salient feature of the Act, wages are not paid on within 15 days of work done and in full amount that they are entitled to. Wages were deducted from each and every beneficiary by the insurgent groups that are more dominant in the region. Even the daily wages of the rural poor are not spared. And moreover delayed payment of wages makes the condition of the beneficiaries at stake. Problem also persists in case of account and records relating to the scheme as there is no transparency.

9.5.iii Issuing of Job Card

Job card is issued to every registered household but it is not issued within 15 days of application. It takes at least a month for the people to receive their job card. And apart from this there is manipulation of job cards by the village council. There is number of fake household created by the village authority to extract more extra income for themselves. Thus, corruption is so immense from the initial stage of implementation.

9.6 Unemployment and MGNREGS among the Youth

The problem of unemployment is the problem of youth unemployment. Large sections of the youth (almost two-thirds) are unemployed. Employment generation like poverty alleviation has also been perceived to be a function of economic growth. Its frailty was not questioned as the debate centred on the fact that India has not been able to achieve the required rate of growth to provide adequate employment. Employment generation was tried to be augmented by special employment generation programmes, such as NREGS etc. The growing realisation about the severity of the unemployment situation and the jobless in the district was lessening with the implementation of an ambitious umbrella programme, the National Rural Employment Guarantee Scheme.

Unemployment is higher for males relative to females, as per the data collected from the sample village. The data is given below:

Employment Pattern of the Selected Households

Attributes	Khongjaron Village		Tamenglong Village		Dailong Village		Total	
	M	F	M	F	M	F	M	F
MGNREGA	30	60	26	64	40	50	96	174
Agriculture & Allied	40	50	30	70	35	65	105	185
Government/private employee	47	16	55	38	25	14	127	68
Business	09	56	13	40	6	44	18	140

Source: Field data.

The results for education across both genders consistently show lower unemployment with literacy going up to initial schooling, and higher unemployment at higher education levels, with variation by gender in terms of whether the change in sign occurs at secondary or higher secondary level. The lower unemployment with education up to primary school is most likely reflective of other correlates of schooling (such as perhaps higher health status of respondents with some schooling) rather than the need for literacy skills on the job.

9.7 Development Programme and Insurgency

The present study reveals the basic facts regarding the development process in Tamenglong District during the preceding decades. Initiatives required to increase the economic development in the district was never been followed up. Meanwhile, insurgency and other problems of a political nature progressively complicated the scenario and governmental attention was consequently diverted to other aspects of the canvas. As a result there has not been any considerable improvement over the years. A study on NREGA in the district reveals that there is upsurge of insurgency in Tamenglong with fractional groups which is responsible for the poor development activities making the life of the people at stake and insecure. It also indicates that there has been an added impoverishment of the rural areas with regard to developmental activity. It is generally seen that insurgency is fuelled by economic backwardness, therefore, has some basis. But it would be totally inappropriate to assume that there is a direct linkage between insurgency and economic backwardness.

The degree of interconnectivity in transportation networks is in a ghastly condition. Production networks characterized by road systems are restrained by rebel groups, since these insurgents group are dependent on taxation or extortion through obstruction. Even in the context of the other indicators of economic and social well being, Tamenglong district is worse off in comparison to other districts of the States. The most concerned aspect is about the disparity of income distribution in the district. In the recent years investors have shunned Tamenglong district, because of militancy and corrupted leaders. It discourages investment and economic development leading to various growing problems like unemployment, which in turn provided a way to join militancy.

The youth of this area looked out for employment by joining various insurgent groups. Since these groups extort money and their living condition is better off than the bureaucrats, in some cases in the district. The populace live in constant fear of raising their voices against this anti-social element in the district. Every developmental programme to be implemented in the district is taken up by the public leaders and the leaders of the insurgent groups. There are more than two insurgent groups dominating the district which gives constant fear and threat to the public. The dominant group takes control of the entire development project allotted to the district. In that case even the MGNREGA which is supposed to be one of the schemes that reached to each and every household is also being targeted. The beneficiaries do not get the exact amount of their daily wages since certain amount of percentage was deducted from it. So, even the hard earn daily wage of poor people are being extorted by the insurgent groups for their own interest instead for the cause.

Insurgency issue is a never-ending discussion in the state. The people of Tamenglong remain cocooned and wish for development of tranquillity with solution to the problems they faced each day. The situation in Tamenglong is such that the people no longer fear the 'outsiders' but rather feared their local factional insurgent groups. Development in the district is hampered due to the violence, extortion, kidnapping, leakage and the sulking and cynical indifference they have engendered to the populace. This has not only delay development but could even deny the peace, progress and prosperity that everybody seeks. Although there are varying estimates, substantial amounts of funds have also been accruing to the insurgents.

In Tamenglong district, the level of unemployment is due to a lack of developmental activity. More than 70 per cent of these are educated as per the 2011

Census. These unemployed youth, to a large extent, also provide the reservoir of manpower from which the various insurgent outfits easily recruit their cadres. Reportedly, even the families of these 'volunteers' do not object because of the lure of financial compensation which, though not handsome (as some sections of media tend to make out), is not meagre either. Furthermore, there also exists the added attraction of the employment and rehabilitation packages in the context of certain ill-considered surrender policies. It needs to be noted that economic backwardness is one of the significant causes fuelling insurgency in the district, although it is not the sole cause and nor can it be said that insurgency is the consequence of the lack of economic development. In the larger canvas of the attempts being made to achieve material progress, it is also true that rampant corruption in the bureaucracy and the political class manning the delivery system stands as a stumbling block as they are insincere in their duties.

The outfits later gained strength primarily through clandestine support networks of likeminded politicians and over-ground front organisations, as also with the ill-gotten monies acquired mainly through extortion. The NSCN, for example, was able to become a large outfit primarily due to the tacit support provided by the politicians, and the public leaders. In such a milieu, insurgency has lowered the income and development, and increased the percentage of the labour force in agriculture. Whether insurgency impacts economic prosperity today? The answer for many is 'yes', although there is much potential mechanism through which insurgency could exert persistent economic development.

Land was a key instrument for generating political support for election. These results are consistent largely with a high level of correlations between land and contemporary economic and political outcomes. But insurgency somehow is responsible for the current economic and political detrimental outcomes, which means that they are partly responsible for the political instability and backwardness in the region. One significant aspect is that most of the developmental schemes that were sanction by the government are captured by these insurgent groups and did not complete the assign task. Bureaucrats, village chiefs, officials and even big contractors lived at their mercy because the killing machines are in the hands of various insurgent groups.

Majority of the youth who are school dropouts or belonging to the BPL are forced to join the insurgent group since they are inspired by the way the insurgent leaders live freely and lavishly from the exploited money. They are the boss and self-style hero of the region, although not welcome by majority of the populace.

Important mechanism linking to insurgency is the increasing level of violence and conflict within the community that threatens the peace and security of the people. The rise of violence leads to instability and reduces economic prosperity which effects many long-term developments. For example, the insurgent group even took control of the educational aspects, health care, road infrastructure in the region where some percentages of almost all the government employees were deducted from their salary. The insurgent leaders also involves in determining the job recruitment for teachers, health care, contractors etc. They decide whom to give the job and whom not to give; who should be transfer and who should be posted in the specific area. On the above basis, the impact of insurgency persists primarily through the district long-run development capacity, which is detrimental to the society.

9.8 Corruption and MGNREGS

Development agenda on various aspects in the district remained a challenged. Safe drinking water, sanitation facilities, poverty reduction remained a challenge in the district. No matter how poor and backward the district is, there is absence of beggars. The most significant strength of the people is their simplicity but now with the implementation of new development programmes in the district it has become complex with the emergence of class in the society where the rich has the upper hand in all the government undertakings especially in securing government jobs.

The scenario in Tamenglong district is such that every department is said to be corrupted in executing their responsibilities and also in recruiting new employees. It is impossible for the poor folk to get a descent govt. job because he has no money though he may be educated and qualified for the job. The tradition is such that the one who can bride with the highest bid to the insurgent and the bureaucrats gets the job. For example, a person who is not that qualified but has money and political connections have better chance in getting the govt. job compared to the most qualified person who have insufficient money to bride for that job. Thus, a rich family is always inclined to buy a higher prestige job.

In Tamenglong, corruption has reached its zenith. Bribery is the talk of a town. It is an open-secret affair of the situation. No matter how qualified a person is, if he has no money and political connections he is likely to remain unemployed. Therefore, the only option for the youth to be engaged himself is to join an insurgent and follow the extortionist culture, instead of living a rightful life. It is the condition that compels him to join, not his dream.

The earlier united simple villagers are now divided on the line of political power due to money and power struggle. Moreover, the traditional authority is being challenged with the implementation of MGNREGS. The village chairmanship which is hereditary among most tribal groups is losing its authority. People do respect and accept this leadership in the past, but in this present context people have lost their confidence in the traditional authority as it failed to adapt to the changes. Thus, corruption is one important element that weakens the traditional system.

These days, there is so much antagonism among the villagers. Development programmes that sanction lots of money has blinded the love for each other. The more one has the political power the more he tries to exploit the weaker section of the people. Recently, in Tamenglong the government initiate in improving the transport system and in this regard the government declares compensation for those families whose land will be affected. Since, money is involved here for the village and private land owners, now the villagers are apprehensive about the role of the village chairman in collecting and distributing the money. Thus, the traditional authority is being challenged by the rich and those in power to bend to their greed wishes. Another conflicting incident occurred between the modern politicians and the traditional authority during the District Election of 2010 and General Election of 2012, where the insurgent groups instead of dissolving the tension amicably took interest in support of the rich and the politicians who bribed them with money.

It is also alleged that there is rampant corruption in the implementation of MGNREGS. The scale and dimensions of MNREGS corruption in Tamenglong suggest that the kind of corruption is impossible without active connivance of the block and district officials. This corruption in many areas has eroded the benefits of the scheme.

Corruption has not spared even Tamenglong one of the most backward districts in the country. Could it be because of high prevailing corruption that the district remains underdeveloped despite various developmental programmes and funds made available to them in the past, or could it be due to geographical location, high illiteracy and traditionalist besides other factors. Land and boundary disputes among family members, lineages, clans and among neighbouring villages are common issues in Tamenglong. Another observation made as to why Tamenglong is still underdeveloped is that most of the development funds meant for poor people are snatched away by the irresponsible officials, potential leaders, insurgents and politicians silently without the knowledge of the public in many cases. Further, once a corrupt leader comes into power, he or she played the hegemony game by appointing only his kins and family friends for different govt. jobs instead of appointing most qualified candidate for the job. Corruption continues to strangle the growth of development including *MGNREGS* despite positive efforts.

9.9 Conclusion

No doubt, despite various predicaments MGNREGS has become a lifeline for many wage seekers of Tamenglong and it has to a significant level improved the living conditions of the masses to certain extent. On the other hand, the *National Rural Employment Guarantee Programme* can be seen both as a curse and a boon for different reasons. In spite of certain government officials making sincere efforts to improve the rural infrastructure of the village, there are certain drawbacks in the NREGS enthusiastic government act, since the village council are not able to play an effective role although its role has been specifically built into the programme. There is also lack of identification of developmental projects. The lacklustre performance has been due to the lack of vertical and horizontal coordination between the multiplicity of implementing agencies comprising of Government at the State and Centre level, and the village authority at the village level. The mechanism of the flow of funds to the village level is highly complicated.

NREGA tries to enable all citizens to reap the benefits and opportunities of development. But many issues have not been articulated by the present government scheme due to massive corruption at the state and grass root level. A large number of opportunities had been set up which have not delineated the path of going ahead to achieve inclusive growth. Thus, not much has been accomplished on the ground which among others requires a complete overall restructure. For this scheme to be inclusive, the policy makers must develop delivery models including a strict accountability system and a collaborative effort in implementing the developmental programmes in the rural areas involving the people to actively participate in the decision making and maintaining transparency for the civil society. Opening bank accounts should be mandatory for each individual in which their wages can be directly transferred to their own accounts.

To sum up, it may be stated that special programmes for employment generation have been rightly accepted by the government as a major need. The scheme has been implemented for a long time now. For over the past years these programmes have not been implemented in an effective manner as expected in Tamenglong district. Each year the government continues to follow the same principle instead of rectifying and changing the implementation process in the desired direction to achieve the targeted objectives. Effectiveness of the scheme can be improved by the involvement of the local bodies like Village council committee, member of district council and the public which have been rightly given a major role in the society.

It is not possible to realise the massive potential of the *National Rural Employment Guarantee Act* if we deploy the same ossified structure in the implementation that has deeply institutionalised corruption, inefficiency and non-accountability into the very fabric of rural society. On the other hand, NREGS is indeed a historic programme as it relief the rural poor to certain extent by alleviating poverty and transforming their livelihood. Overall the impact is quite acknowledgeable. On the other hand, if necessary reforms are put into right place, NREGA will hold out the prospect of transforming the livelihoods of the poorest and herald a new era towards rural development.

References

Chambers, Robert. 1983. *Rural Development: Putting the Last First*. London: Longman.

Bottomore, T. B. 1986. *Sociology- A guide to problems and literature*. Bombay: Blackie and Son (India), Ltd.

Kumar, D.V.2008. 'Engaging with Modernity: Need for a Critical Negotiation' in *Sociological Bulletin*. Vol.57. No.2 (242-254).

Mishra S.N & Kaushal Sharma. 1983. *Problems and prospects of R.D. in India*. Delhi: Uppal Publishing House.

Moodithaya M. S. Shetty N. S., Thingalaya N. K. 2011. 'Nrega: Employment Guarantee Programme and Pro-Poor Growth' in *The Study of a Village in Gujarat*. New Delhi: Academic Foundation.

Mohanty, T.P. 2010. *Rural Development and National Rural Employment Guarantee Act*. New Delhi: Axis Publications.

Oommen, T.K. 2004. *Development Discourse: Issues and concerns*. New Delhi: Regency Publications.

Pandey.1985. *Sociology of Development: Concepts Theories and Issues*. New Delhi: Mittal Publication.

Sachs, W.1994. 'Introduction' in W Sachs (ed). *The development dictionary: A guide to power*. London and New Jersey: Zed Book Ltd. (pp. 1-5).

Seers, Dudley. 1972. *What are we trying to measure?* Brighton Institute of development Study: IDS reprints.

Willis, Katie. 2005. *Theories and practices of Development*. London and New York: Routledge.

About the Author

Pr. Dimchuiliu is currently pursuing her Ph.D. in the Department of Sociology, NEHU, Shillong. She also works as part-time lecturer in Women's College, Shillong.

E-mail: dimchuiliu@gmail.com   Phone: 8794941927

Chapter 10

Grassroots Democracy under Autonomous District Council in Assam: Problems and Prospects

☆ *Robert Tuolor*

ABSTRACT

The 73rd and the 74th Constitutional amendment acts, which paved the way for three tier Panchayati Raj institution and municipal bodies, are not applicable in the two hill district of Assam, viz: The North Cahar Hills district and Karbi Anglong District, since these two district are listed in the Sixth Schedule of the Constitution. An alternate structure of local-self Government, called the Autonomous District Council operates here. The Autonomous District Councils not only give the hill people of North-east India training on local self government but also bring in faster economic development by associating people with the developmental works through their representatives in the District Councils. The Autonomous District Councils have been empowered to enact legislations for the welfare and development of the hill tribal people of the North-east. The Autonomous District Council may also be described as a 'State in miniature' having all the paraphernalia of a government like Legislature, Executive and Judiciary. It has full autonomy to legislate and administer on subjects like Land Revenue, Primary Education, Customary Laws etc. assigned to it under the Sixth Schedule of the Constitution of India.

Here the study mainly focuses on the N.C. Hills Autonomous Council of Assam. The study examines the working of this Autonomous Council and highlights the problem and prospect in providing grassroots democracy to the hilly people.

10.1 Introduction

Democracy is never complete unless active involvement and participation of the people at all levels is assured. Particularly in the modern age of democracies, where it is said that the state, the Government and power belong to the people: the Government is the servant and welfare agency of the people. The people come in contact with the Government only at the local level. Through local Government institutions the new political elite make an attempt to mobilize the mass for the nation-building task of economic and social development. (Narang 2000:285) Self-governing institution are integral and indispensable part of the democratic process. "Grassroots of democracy" based on small units of government enables people to feel a sense of responsibility and to inculcate the values of democracy and at the same time it offers a unique opportunity to participate in public affairs, including developmental works. Self governing rural local bodies are described in the Indian context as institutions of democratic decentralization or Panchayati Raj (Fadia 2005:561). Under the British rule, the *Panchayat* gradually disappeared because of the compelling need of the foreign ruler to centralize. Though Lord Ripon a British Viceroy took the initiative of establishing an alternative system of local-self government in India the institutions that came into being then lacked in resources as well as autonomy. In the twentieth century, Gandhiji openly articulated the need to revive the *Panchayats* with democratic bases of their own and invest them with adequate powers so that the villagers could have a real sense of *swaraj* (self-governance), (Palanithurai 2004:109). Finally, the 73rd Amendment Act 1992 which was brought into effect from April 24, 1993 clearly envisages the establishment of *Panchayats* as units of local-self government. It empowers them to prepare plans for economic development and social justice and also implement schemes in these areas as may be entrusted to them by the respective state governments.

In Assam, a new Panchayat Act was enacted in 1948. It provided for the division of rural Assam into Panchayat areas, with each area consisting of a number of villages and each village having a primary Panchayat. A Government committee, appointed in 1953 had recommended that Panchayat should be established throughout rural Assam within a period of five years. A few years later, the Assam Panchayat Act. 1959 was enacted. This Act, enforced from 1960 provided for a three-tier system. Thus after some initial setbacks democratic decentralization as recommended by the *Balwant Rai Mehta Committee* was introduced in Assam (Institute of Social Sciences 2000:70).

However, the 73rd and the 74th Constitutional Amendment Acts which paved the way for three tier Panchayati Raj institution and municipal bodies are not applicable in the two hill districts of Assam *viz.*; The North Cahar Hills district and Karbi Anglong district, since these two districts are listed in the Sixth Schedule of the Constitution of India. An alternate structure of local-self Government called the *Autonomous District Council* operates here. The idea behind the Sixth Schedule was to provide the tribal people of North-East India with a simple administrative set up which can safeguard their customs and ways of life and to provide autonomy in the management of the affairs. The Autonomous District Councils not only give the hill people of North-east India training on local self government but also bring in faster economic development by associating people with the developmental works through their representatives in the District Council (Dutta 1999:256). The Autonomous District Councils have been empowered to enact legislations for the welfare and development of the hill tribal people of the North-east. The Autonomous District Council may also be described as a 'State in miniature' having all the paraphernalia of a government like Legislature, Executive and Judiciary. It has full autonomy to legislate and administer on subjects like Land Revenue, Primary Education, Customary Laws etc. assigned to it under the Sixth Schedule of the Constitution of India.

Here the study mainly focuses on the N.C. Hills Autonomous Council of Assam. Its main objective is to study the working of the N.C. Hills Autonomous Council to examine whether it succeeds in providing grassroots democracy to the people of the district. The paper also investigates the limitations, if any and suggests measures to rectify the problems.

The North Cachar Hills district was created in 1867 under the province of Assam and shortly afterward it was tagged to the Cachar district as its sub-division (Dutta and R. Bhuyan 2007:626). Under the Government of India Act 1935, the N.C. Hills district was classified as 'Excluded Areas' and was administered by the Governor himself in his discretion where the ministers had no constitutional right to advise him in connection with its administration and no act of the federal or provincial legislature would apply to this area unless the Governor applied it with some exception or modifications. The Cabinet Mission in May 1946 suggested that there should be an advisory committee on the rights of citizens, minorities, tribals and excluded areas (Dutta 1993:7). Accordingly, after independence the Constituent Assembly of India set up an advisory committee and this committee appointed a sub-committee generally known as *Bordoloi Committee*. The recommendations of the sub-committee were incorporated in the Sixth Schedule to the Constitution of India (Rao 1976:149). Since its amalgamation with the district of Cachar, North Cachar Hill was a sub-division of that district and this is how the name of the district came to be known as North Cachar Hills district (Dutta and Bhuyan 2007:627). Later in 1951, it was amalgamated with Mikir Hills (present Karbi Anglong district) and formed a separate civil district known as United Mikir and North Cachar Hills district. The N.C. Hills district remained a sub-division of that district until February 2, 1970 on which date it attained its present status as a full-fledged autonomous district to be administered under Sixth Schedule of the Constitution (Tapadar 2007:7).

This chapter examines the Legislative, Executive and Judicial organs of the N.C. Hills Autonomous Council from 1995 to 2008 since additional powers were given in the year 1995 under the amendment of the Sixth Schedule of the Constitution.

10.2 Power and Function of the North Cachar Hills Autonomous Council

Legislative Function

Regarding legislation, Paragraph 3 of the Sixth Schedule to the constitution of India gives the North Cachar Hills Autonomous Council the power to make laws.

The study reveals that during the period of study, the Autonomous Council of North Cachar Hills had passed six amendments to the Principal Acts concerning more on the constitution of the District Council. Regarding rule making, not much had been done by the North Cachar Hills Autonomous Council excepting on Contributory Provident Fund and the rules on trading by non-tribals.

Executive Function

The Executive functions of the North Cachar Hills Autonomous Council may be sub-divided into different departments but here the discussions will mainly be confined with special reference to the administration of Land Revenue, Forest, Taxes, Finance, Administration of villages and towns and other subject including administration of justice relating to social customs of the tribal people of the district.

Land Revenue Administration

Under paragraph 8 of the Sixth Schedule the Autonomous Council is given the power to assess and collect land revenue in respect of lands under the jurisdiction of the district. Land Revenue is a major source of income for the Autonomous Council. The Land Revenue assessed during the period between 1995 and 2008 was Rs. 74,65,328,30. The Lands under permanent paddy cultivation and permanent houses in Haflong town and market areas are assessed to land revenue. *Jhum* land (Land used for shifting cultivation by the hill tribals in North-east India) and other village lands are not assessed to land revenue as there is no fixity of cultivation, and hence it was difficult to access these lands to land revenue. Moreover, the unfavorable economic condition of the people as well as food scarcity caused by drought, flood etc, irregular collection of land revenue due to ethnic conflicts in the district and deplorable communication make the collection of revenue low and difficult.

Forest Administration

According to paragraph 3(b) of the Sixth Schedule to the Constitution of India, the North Cachar Hills Autonomous Council is empowered to manage any forest other than a reserved forest within the district. Revenue collected from the sale of various forest products goes to the exchequer of the Autonomous Council. Forest is the main source of the council's income. A good percentage of its income comes from forest. During the period of study, the revenue collected in the year 1995-96 was the highest which is Rs.4,22,57,456.00 but declined drastically to Rs. 2,27,54,643.00 in 2008. The year 2002-03 shows the lowest ever collected which is Rs.73,02,284.00. To sum up, the revenue collected during the period of study kept on declining due to depletion of forest cover, depletion of forest resources and degradation of land as a result of *jhum* cultivation coupled with illegal felling and deforestation for creating agricultural land. The Supreme Court's ban on cutting of timbers further reduced the Council's income from the Forest department.

Taxes

As per the provisions under paragraph 8 of the Sixth Schedule to the Constitution of India, the North Cachar Hills Autonomous Council levies and collects land revenue, forest royalty, taxes on professions, trades, and employment, taxes on vehicles, animals and boats, taxes on entry of goods into market for sale and tolls on persons and goods carried in ferries. The Autonomous Council also collects market taxes, comprising of rent on shop sites, latrine tax, license fees, water tax and sales proceeds. The study shows that tax is one the main source of the Council internal income but taxes was not collected regularly and properly due to some loopholes and leakages in the process of collection.

Finance

The Sixth Schedule of the Constitution of India in its Paragraph 13 mentions that the estimated receipts and expenditure pertaining to an Autonomous District which are to be credited to, or is to made from the Consolidated fund of the State will be first placed before the District Council for discussion and then after such discussion be shown separately in the annual financial statement of the State and is to be laid before the legislature of the State. (Bakshi 2005:354).

In Assam, the North Cachar Hills Autonomous Council prepares the estimated receipts and expenditure pertaining to N.C. Hills Autonomous Council in respect of entrusted subjects/ Departments as per the allocations, norms, etc. communicated by the State Government, and in respect of Non-entrusted subjects/ Departments the State Government prepares the estimated receipts and expenditure of the N.C. Hills Autonomous Council and forward the same to it.

The N.C. Hills Autonomous Council in their council session considers the estimated receipts and expenditure pertaining to N.C. Hills Autonomous Council in respect of both entrusted and non-entrusted subjects/ Departments and sends the same along with the synopsis of their consideration to the State Government within the stipulated time. Ordinarily no changes will be made in the estimated receipts and expenditures as considered by the N.C. Hills Autonomous Council in respect of the entrusted subjects/ Departments. However, if it becomes necessary to do so, the council will be consulted and reasons for the proposed changes will be explained. The State Government then places the estimated receipts and expenditures as considered by the N.C. Hills Autonomous Council along with the synopsis before the State Legislature.

The North Cachar Hills Autonomous District Council derives its income mainly from two sources namely; internal (from its own sources) and external (from the grant-in-aid from the State Government). The major sources of internal revenue of the North Cachar Hills Autonomous Council are royalty from forest, land revenue, taxation on profession, trades, callings, employment, vehicles and miscellaneous receipts. The total internal revenue of the Council during the period of study was Rs. 142,29,47,700/-. The only external source of income of the Council was the Grant-in-aid from the Assam Government amounting to Rs.256,12,88,653/-. The percentage of the total internal income of the Autonomous Council to that of the grants from the Government of Assam during 1995-2008 were only 35.71 per cent . Thus, during the period of study 64.29 per cent of the receipt of the Council came from the Assam Government in the form of grants.

10.3 Judicial Function

As per Paragraph 4 and 5 of the Sixth Schedule to the Constitution of India, The North Cachar Hills Autonomous Council is given the powers of constituting Village Courts or Council and other Appellate Courts for trial of suits and cases between the parties, all of whom belonging to the Schedule Tribes of the district. It may appoint suitable persons to be member of such Village Courts or Presiding Officer of such courts and may also appoint officers as when necessary for the administration of the laws made under the legislative powers. It exercises the power of a court of appeal in respect of all suits and cases triable by Village Councils or Courts constituted, and no other courts except the High Court of Assam and the Supreme Court have the jurisdiction over such suits and cases.

The number of cases tried by the North Cachar Hills Autonomous Council during the period from 1995-2008 are 20 land disputes, 70 title suits and 10 miscellaneous cases. The Sixth Schedule to the Constitution of India has also provided in Paragraph 3 that the North Cachar Hills Autonomous Council has the power to make laws in respect to the establishment of Village or Town Committee or Council, their powers and any other matter relating to village or town administration including village and town police and public health and sanitation. However, the Village Courts or Village Councils as provided in paragraph 4 of the Sixth Schedule to the Constitution has not yet been constituted by the North Cachar Hills Autonomous Council. However there exists the traditional village Panchayat in all the villages of the district. These village Panchayats are constituted with the village elders generally presided over by the *Mauzadars* (tax collector in rural areas or the village headman) who has been authorized to hear disputes concerning tribal laws, customs and also cases of petty civil and criminal natures. These Panchayats still decide petty cases in their respective villages according to their customs and traditions. These Panchayat, however are not to be confused with the Panchayat elected under the Assam Panchayat Act 1959 nor with the Village Council which the Autonomous Council may have set up under paragraph 4 of the Sixth Schedule to the Constitution of India. These were in existence even before the North Cachar Hills Autonomous Council came into being. The organization of these village Panchayats differs from tribe to tribe. The Chief Village authorities of these traditional Panchayat such as *Mauzadars* and *Gaonburas* (village headman) are authorized to perform the functions and duties of the Village Courts. The tribal people, in general in the North Cachar Hills district like any other hill tribe are not litigious, and almost all the disputes are settled at the village levels. Most of the cases that came up to the Subordinate District Council Court also eventually ended in a compromise.

10.4 Conclusion

An analysis of the functions of N.C. Hills Autonomous Council reveals that in spite of the 1995 amendment, the Autonomous Council has been laden with innumerable hardships which make it difficult to provide grassroots democracy to the people. N.C. Hills district is one of the most backward districts in Assam with no industries or factories. Although the district is one of the biggest district in Assam in terms of area with some natural resources but absence of advanced technology to exploit them makes it difficult to earn any revenue. Under such circumstances, the Council could not run smoothly with almost any income of its own. It has to depend mostly on the grants received from the State Government to meet its daily requirements. The N.C. Hills Autonomous Council also suffers due to its own incompetence and reluctance. A series of mal-administration is evident in its functioning like irregular collection of land revenue, depletion of forest resources and degradation of land as a result of jhum cultivation, illegal felling of trees and deforestation for creating agricultural land. Apart from that, some of the provisions of 1995 amendment are not implemented wholly. There are also instances of interferences by the State government in many occasions and the funds meant for the N.C. Hills Autonomous Council are not released on time. Other important hurdles that impede the functioning of grassroots democracy in N.C. Hills district are militancy and corruptions. Under such circumstances, the N.C. Hills Autonomous Council, which is the local-self governing institution could not fully succeed in providing grassroots democracy to the hill tribal of Assam.

References

Bakshi P.M. 2005. *The Constitution of India*. New Delhi: Universal Law Publishing Co. Pvt. Ltd.

Chaube S.K. 1973. *Hills Politics in North-East India*. Calcutta: Orient Longman Ltd.

Dutta Ray. B. 1999. Autonomous District Council and Strategy of Development in North-East India, in Amalesh Banerjee & Biman Kar (eds). *Economic Planning and Development of North-East States.* New Delhi: Kaniska Publishers.

Dutta and R. Bhuyan, (eds). 2007. *Genesis of Conflict and Peace: Understanding North-east India: Views and Review.* Vol.2. New Delhi: Akansha Publishing House.

Dutt K.N. 1979. *Assam District Gazetteers*. Guwahati: United Mikir and North Cachar Hills District. Government of Assam.

Fadia B.L. 2005. *Indian Government and Politics*. Agra: Sahitya Bhawan Publications.

Hansaria B.L. 1993. *Sixth Schedule to the Constitution of India-A Study*. Guwahati: Ashok Publishing House.

Meenakshisundaram S.S. 2004. Rural Development as a Mechanism to Strengthen Panchayati Raj, in G. Palanithurai (eds.). *Dynamics of new Panchayati Raj System in India*. Vol.III. New Delhi: Concept Publishing Company.

Narang A.S. 2000. *Indian Government and Politics*. 6th Edition. New Delhi: Gitanjali Publishing House.

Souvenir. 1999. North Cachar Hills Autonomous Council 48th Foundation Day Celebration. 29th April, N.C. Hills Autonomous Council. Haflong: Cultural & Publicity Dept.

Souvenir. 2008. The 57th Foundation Day. 29th April. North Cachar Hills, Haflong: Dept. of Information and Public Relation.

The Constitution of India. 2001. Lucknow: Eastern Book Agency.

Tapadar A Zahid. 2007. *N.C. Hills. The Paradise of Exploration.* Haflong: Dept. of Information and Public Relation.

Venkata V Rao. 1976. *A Century of Tribal Politics in North-east India* (1874-1974). New Delhi: S. Chand & Company.

Some Primary Sources Gathered during the Field Work

1. Records collected from the Legislative branch of the N.C. Hills Autonomous Council, Haflong, Assam.

2. The North Cachar Hills District Council Employees' Contributory Provident Fund (Sixth Amendment) Rules, 1998.

3. Records collected from the Land Revenue Department, North Cachar Hills Autonomous Council, Haflong.

4. Records collected from the Forest Department, Divisional Forest Office, N.C. Hills Autonomous Council, Haflong.

5. Records collected from the Finance Department, N.C. Hills Autonomous Council, Haflong, Assam.

6. Records collected from the judiciary branch of the North Cachar Hills Autonomous Council, Haflong.

7. *Office Memorandum*, No. HAD/57/95/309, Dated Dispur, the 31st December 1996, Government of Assam, Hill Areas Department.

8. The Assam Autonomous District (Constitution of District Council) Rules, 1951.

About the Author

Robert Tuolor is a Research Scholar in the Department of Political Science, North-eastern Hill University (NEHU), Shillong- 22. He was recently awarded his Doctorate Degree.

E-mail: robert_tuolor@yahoo.com   Phone: 07308997718

Chapter 11

Intricacies of Autonomous District Council of Manipur

☆ *P.G. Jangamlung Richard*

ABSTRACT

In accordance with the provision of Manipur (Hill Areas) District Council Act 1971 enacted by parliament under 5th Schedule of Indian constitution, the government of Manipur instituted Manipur (Hill Areas) District Councils (ADC) in 1973. However, due to the agitations by tribal demanding for 6th Schedule in hill areas the Manipur (Hill Areas) District Council got defunct from 1983 till 2010. Autonomous District Councils with several amendments re-introduced in 2010 after been defunct for more than two decades yet could not perform its functions efficiently. It can be observed from the successive amendments and demands put forward by tribal that there is a clash of interest between Government of Manipur and tribal which led unhealthy politics. Thus in the light of this controversy, the paper attempts to briefly discuss the modern concept of autonomy, the provision given under 5th schedule in the constitution and the changing pattern of powers, functions and limitation of Autonomous District Council of Manipur.

Keywords: *Autonomy, Schedule tribe, Naga, Kuki, Rights and Equality.*

11.1 Introduction

Manipur is a multi-ethnic state with 34.20 percent of schedule tribe, out of total populations of 20,294,000 (Indian Census 2001). Tribals in Manipur may be broadly divided into 2 major ethnic groups *i.e.* Kuki and Naga. Further these two ethnic groups are subdivided into 33 recognized Scheduled Tribes of India (Kamei 2008:175).

Each sub-tribe has distinct dialect, and many tribes have peculiar character of linguistic variation from village to village that they cannot understand even among the same tribe. They have different culture, customs and traditions. Their traditional political system is different from the mainstream Indian political system. It has village boundary, regulated land ownership, a judicial system and social institutions base on customary laws (ibid). Since the inception of the Indian constitution the importance of this tribal traditional political system was recognized and endorsed by the framer.

As recommended by Bordoloo Committee certain provisions were added to the Indian constitution in the 5th Scheduled and 6th Scheduled to protect traditions and safeguard the interest of the tribal in the North-east. The hill tribal in Manipur came under the 5th schedule. Accordingly in 1971, the Manipur (Hill Areas) District Council Act was passed in Parliament. Later District Councils was adopted by Government of Manipur and the first Autonomous District Council election was conducted in six districts in 1973. However, twenty years of experienced (1973-1993) with ADC under 5th scheduled failed to address the aspirations and interests of the tribal. As a consequence, the tribal of Manipur started demanding for 6th Scheduled which was granted to other hill tribes in North-east India. It was followed by protests and demonstrations by the hill tribal leading to total rejection of ADC of Manipur in 1993. Hence, it remained defunct till 2010.

The ADC of Manipur was re-introduced in 2010 after two decades with an assurance from Government of Manipur to devolve more power and functions to ADC. After having implemented for about 20 years and altered by several amendments with the hope of improvements it failed to function efficiently. However, there are many limitations even in the present ADC that crippled to perform its function. It was flatly rejected by *United Naga Council* (UNC) sticking to the demand of provision under 6th Scheduled and the majority of the Naga did not participate in the latest 2010 election. The elected members of the ADC also expressed their grievances in the form strike and demonstration. The ADC members' organization sent memorandums to the state and the Central government demanding for devolution of power and funds. Suspicious and chaos on the tribal due to inefficient and limited functioning of ADC it led to unhealthy politics at the grass root level in the areas dominated by tribals in Manipur. For better understanding, a brief analysis on the concept of autonomy and the provisions under the 5th Scheduled may be discussed.

11.2 Concept of Autonomy: A Theoretical Framework

The concept of autonomy is contested and debated with much controversy. The meaning of autonomy in general refers to lead one's own rational decision without irrational influences whether externally or internally. It has been using as a basis in many human arenas such as ethic, legal, right, policy making etc. The term Autonomy is derived from Greek words 'auto-nomos'; auto means 'Self' and nomos stands for 'law' which literally refers to one who gives oneself their own law or lives under one's own rule (Wikipedia 2013).

In political perspective, autonomy has its own connotation. Its concept was evolved after the achievement of human enlightenment, where men were no longer ready to accept the idea of paternalistic intervention. Its origin is derived from the ancient Greek philosophical idea of 'Self-mastery'. It was indirectly influenced by the Plato and Aristotle idea self-sufficiency (*Auterkeia*) and rule by reasons (just soul). According to their philosophy, Self-sufficiency (*Auterkeia*) without depending on others and rule by human rational soul *i.e.* reason the best condition for happiness.[1]

The concept of autonomy draws nearer in the modern concept to the Rousseau's idea of moral liberty. According to Rousseau the freedom of the individual is important, however for freedom individual is necessary to be in larger community. According to him, human beings are pulled in two directions *i.e.* human being of their own self and on the other hand integrate themselves into a community which is neither mutually exclusive nor easily reconcilable. It is a moral process whereby individuals consciously integrate themselves into a community where everybody is equal and such community is where individuals are best protected from each other rather than trying to eradicate the tension between individuals (Froese 2001:579). Rousseau affirms that the effort to reconcile this tension is the foundation of morality (ibid.:581). The community of equal, Rousseau talked about is formed by individual participating in making its laws. Thus, where individual think as members rather than as individuality that are merely protected by community. Individual freedom and participation in the Rousseau's community of equal is a kind of political autonomy.

Emmanuel Kant's idea of having 'self will', an authority over one's actions in his moral philosophy has been using as a basis of modern concept of autonomy. The self-will according Kant is to determine its guiding principles instead of being obedient to an externally imposed law or religious precepts. He professed individual own understanding according to what he called own guided maxims without the direction of other. He draws a categorical principle in his moral philosophy "Act always according to the maxim whose universality as a law you can at the same time will" (Kant 1785:42). He also says that 'always act in such a way that you and any rational being can accept such will that this dictum of our action should become a universal law' and 'all rational being are treated equally not merely as a means but as an ends in a community and such systematic union of rational being through common objective laws he calls 'Kingdom of ends" (Kant 1785:39). In Kant's kingdom of end, rational being is law givers in their own will where all are treated equal and as an ends. Thus according Kant, autonomy in the political context deals with the right of a person to any action in accordance with universal law. J.S. Mill's gave similar views, emphasizing the rights of individuals to pursue their own goals opposing paternalistic interference except "The only part of the conduct of anyone for which he is amenable to society is that which concerns others" (1859:13). In the part which merely concerns himself, his independence, of right, over himself, his own body and mind, the individual is sovereign.

In addition, political autonomy also deals with the limitation when it comes to the question of interfering other rights or autonomy. However, paternalistic intervention is allow with justification and thus autonomy plays a role bar or yardstick to measure the limitation (John 2004). Non-interference is generally seen as a key to political autonomy. Gerald Gaus specifies that "the fundamental liberal principle" is "that all interferences with action stand in need of justification" (2005:272). Thus in short, modern concept of autonomy in political perspective can be seen as right to pursue one's interests without undue restriction and non-interference.

11.3 The Autonomy and the 5th Schedule Provision of the Indian Constitution

In the context of Autonomous District Council of Manipur, its theoretical concept is drawn from the 5th Schedule of the Indian Constitution. In the articles 244(1) of the Indian constitution provides special provisions for the administration of 'Schedule Areas', in certain areas within a State or Union Territory (Basu 2005:285). Further the 5th Schedule also provides that the tribes Advisory Councils are to be constituted to give advice on welfare and advancement. The system of administration in this provision provides executive power of Union Government to give directions to the respective states related to the administration of these Schedule Areas. The concerned Governor of such state having Schedule Areas have to submit administration reports annually or on the demand of president. The concerned governor also authorized to direct any particular act of parliament or legislature of the states shall apply to the schedule areas or shall apply only subject to exceptions or modification etc (ibid.). This special provision is given to protect and safeguard the interest of the tribal and enhance speedy development of backward tribal living in Hill Areas of Manipur.

11.4 Changing Pattern of Powers and Functions of Manipur Autonomous District Council

In accordance with the 5th Schedule of Indian constitution, Parliament enacted the Manipur (Hill Areas) District Council Act in 1971 to establish Autonomous District Councils in the Hill Areas in the then Union Territory of Manipur (Hill Areas District Council, Act, 1971, Act No: 76). In the year 1972, the Government of Manipur adopted Manipur (Hill Areas) District Council Act 1971 to introduce Autonomous District Councils (ADC) in the five hill districts. They are; 1) Manipur North Autonomous District Councils, 2) Sadar Hills Autonomous District Councils, 3) Manipur East Autonomous District Councils, 4) Tengnoupal Autonomous District Councils, 5) Manipur South Autonomous District Councils and 6) Manipur West Autonomous District Councils. The ADCs were empowered to make recommendations to the state government to bring legislation on matters concerning the members of the Scheduled Tribe on the following matters:

   a. Appointment or succession of chiefs,
   b. Inheritance of property,
   c. Marriage & divorce and,
   d. Social customs.

However, the provisions of Manipur (Hill Areas) District Council Act (1971) were found to be inadequate to safeguard the rights and interest of the multi-ethnic Hills people. Therefore, first minor Amendment was made by the State Government in 1975, but the same could not bring tangible results for the Hills people. Hence, the hill tribal started demanding for extension of Sixth Schedule to hill districts of

Manipur. In 1993 it was followed by protest and demonstrations leading to total rejection of ADC of Manipur. As a consequence the Government of Manipur has dissolved all the ADCs in 1993.

The Government of Manipur enacted Autonomous District Council Act, 2000 by repealing the 1971 Act (Manipur Hill Areas District Council Act, 2000). However the 2000 Act could not be implemented as the tribal had rejected it. In the year 2006 the Manipur government repealed Autonomous District Council Act 2000 and replaced it with Manipur (Hill Areas) District Council (Second Amendment) Act 2006. This Second Amendment of 2006 also failed to appease the tribal. In 2008 the Governor of Manipur promulgated the Manipur (Hill Areas) District Council (Third Amendment) Ordinance exercising his powers provided under Article 213 of the Constitution of India. The District Council Delimitation Committee was also constituted by the Governor to expedite the process of re-institution of ADCs. Thereafter the State Government enacted the Manipur (Hill Areas) District Council (Third Amendment) Act, in October 2008.

The third Amendment Act of 2008 brought some minor changes. The strength of the ADCs increased from 18 to 24. The administrative functions increase from the existing 17 to 26 by inserting a few more areas like fisheries, co-operative, sports & youth affairs, adult & non-formal education, horticulture and floriculture, rural housing scheme, village and cottage industries, small-scale industries, non-conventional energy sources, library and culture activities (Kartika 1930:4,5). The ADC was given the power to recommend to the State for recognition of villages subject to resolution passed by a simple majority of the Council.

11.5 Problems and Challenges

The ADC in Manipur created under the Act of 1971 is different from the ADC under the Sixth Schedule of the Constitution though they are part of North-East hill tribe of India. The executive, legislative and judicial powers of Autonomous Council in other North-Eastern States are basically drawn from the provision of Sixth Schedule of the Constitution of India. However, ADC of Manipur is empowered with only limited administrative powers as incorporated in the Manipur Hill areas District Council Act 1971, without legislative or judicial powers. Even the first Amendment Acts of 2000, second Amendment Acts of 2006 and the Third Amendment Acts of 2008 have consecutively failed to address the aspirations and demand of the Manipur tribals. These Acts does not provide sufficient legislative and judicial powers to the ADC of Manipur. The ADC is not empowered to generate its own revenue. It is dependent on grant-in-aid from the state government. Since the first election in 1973 this limited power of ADC to Manipur could not function effectively to meet the needs of the Hills people.

Despite the rhetoric of safeguarding the rights and interest of the Hills people given in the Constitution, the ADCs failed to deliver governance. The tribal people of Manipur are against the ADC (L. Shangreiso 2010; Somi 2010; Kamei 2000:21; and Bhatia 2010). They have been demanding for ADCs the status under the 6th Scheduled that provides and have the privilege of exercising legislative, executive and judicial powers to protect and safeguard the interest of the tribal people. They continue to demand to restore their tribal autonomy, to safeguard social practices and customary laws, to protect the interest and rights of the tribal over their land etc. Though the government of Manipur has tried to appease through various amendments of the 1971 Act, yet the tribal have their own reasons of suspicions and apprehensions since even 2008 Act also failed to address the basic demands of the tribal.

To support the evident facts, one could see the drastic failures of ADC. Thus, the framers of the Constitution specially designed the 5th and 6th Scheduled as a mechanism to protect the interest of Tribal by giving political autonomy, and to enhance speedy development through additional "fund" from the Center for the councils to execute their works effectively. The district councils in Manipur under the Act of 1971 have proved ineffective to bring any remarkable development in the hill areas during its functioning for nearly 20 years since 1972. This proves that ADC is formulated and implemented as an institution without judiciary and legislative power. It fund is placed at the mercy of the state government without any well defined source of income. Hence the tribal irrespective of community oppose the ADC as it cannot address their aspirations and problems.

Though the latest amendment Act 2008 was proclaimed to enhance the ADC's powers and functions in the Manipur State it brought only some minor changes, instead of incorporating the demands of the tribal. It incorporates only those items limited to maintenance, repair and management which have nothing to do to safeguard the tribal interest or judiciary, legislative and financial autonomy. Even today the provisions given in the 2008 Act is yet to be implemented successfully. The elected Council members are kept at mercy of the ruling party.

Here, one can argue that the powers and functions of ADC of Manipur cannot be called Autonomous as discussed above. Autonomy in political perspective defines as right to pursue one's interests without undue restriction and non-interference. The present situation of ADC of Manipur has no power even to implement those provisions given in the constitution due to limited devolution of power. Over and above there are two layers interferences over the ADC of Manipur *i.e.* Central government through Governor and State Government over the ADC of Manipur. Thus it can also be said that autonomy empowered under the Districts Council Acts is not to the standard of modern concept autonomy.

Here again, one can argue that there is discrimination in the constitution itself to hills tribals of Manipur by providing 5th Scheduled who are not different from other North-eastern tribal in every aspect. They are discriminated from enjoying the facilities provided under 6th Scheduled and to exercise the provisions of executive, legislature and judicial powers.

11.6 Conclusion

ADC as an institution to empower the tribal to safeguard their interest and to effectively participate in the developmental process in Manipur has turned out to be politically contentious issue between the majority non-tribal who dominate the Manipur government and the marginalized hill tribal who are being treated as puppets to the government. It may be observed from above successive amendments and demands put forward by tribal that there is a clash of interest between the people of valley and hills. The tribal view the various steps taken by government of Manipur as tactical move to annihilate the tribal and their land. There is also a big question as to why the government of India and particularly Manipur goes on discriminating the tribals of Manipur. They faced the same situation and problems like any other tribal of North-east India however, the constitution seems treating the Manipur tribal unequally under the provision of safeguarding their own political interest. Therefore, it may be concluded that the inefficiency, dysfunction and chaotic problems of Manipur ADC is not only limited to majority and minority interest clashes within the state but is also found unconstitutionally sound with discrimination. Thus, the question of fair and equal treatment for all the tribals is very much neglected in the case of Manipur ADC, which need a serious retrospection if it is to revive and made functional again with regards to the 5th and 6th Scheduled provisions laid out in constitution.

Endnotes

According to Plato the human soul has three parts: reason, spirit (the source of moral indignation) and appetite (stomach). If spirit and appetite overcomes the prudence of reason bad results follow, harmony and happiness prevails when reason subdue the lower other two human souls (Mackenzie (1985: 89).

References

Autonomy, Accessed on 12-01-2013, http://en.wikipedia.org/wiki/Autonomy

Basu, Durga Das. 2005. *Introduction to the Constitution of India.* (19th edition). New Delhi: Wadhwa and Company Laws Publishers.

Bhatia, Bela. 2010. 'Justice Denied to Tribals in the Hill Districts of Manipur' in *Sanhati,* 12th July 2010. Accessed on 24-04-2011, http://sanhati.com/excerpted/2546.

Census of India. 2001, Accessed on 26-09-2011, http://censusindia.gov.in/Tables_Published/SCS T/ dh _ stmanipur. pdf.

Christman, John. 2004. 'Relational Autonomy, Liberal Individualism, and the Social Constitution of Selves', in *Philosophical Studies.* Vol.117, No. 1-2. (pp. 143,164).

Froese, Katrin. 2001. 'Beyond Liberalism: The Moral Community of Rousseau's Social Contract', in *Canadian Journal of Political Science / Revue canadienne de science politique.* Vol. 34, No. 3, Sept. (pp. 579, 581).

Gaus, Gerald F. 2005. 'The Place of Autonomy within Liberalism', in *Autonomy and the Challenges to Liberalism.* (pp. 272-306).

Kamei, Gangmumei. 2008. *Ethnicity and Social Changes.* New Delhi: Akansha Publishing House.

Kamei, Gangmumei. 2000. 'Village Administration in the Hill Areas of Manipur', in *Ethnicity and Social Changes.* Vol.4, No.9. Sept.

Kant, Immanual. 1785. *Grounding for the Metaphysics of Morals*, Translated by James W. Ellington, (3rd edition.). Indianapolis: Hackett Publishing Company,Inc.

Kartika (5). 1930. The Manipur (Hill Areas) District Councils (Third Amendment) Act, 2008 (Manipur Act No. 7 of 2008), No. 276 (A) Imphal, Monday, 27 October 2008, (P.4).

Manipur (Hill Areas District Council) Act, 1971, Act No: 76, 26th December, 1971

Manipur Hill Areas District Council Act, 2000.

Manipur Hill Areas District Council Act, 2006.

Manipur Hill Areas District Council Act, 2008.

Mill, John Stuart. 1956. *On Liberty.* Indianapolis and New York: The Liberal Arts Press.

Shangreiso, H. L. 2010. 'Divided Manipur: Autonomous District Council Act', in *North-east Herald.* July 2010, Accessed 28-04-2011, http://north-east-herald.blog spot.com/2010 /07/divided-manipur-autonomous-district.html.

Shimray, UA. 2009. 'Land Ownership System of Naga Community: Uniqueness in Tradition', in Priyoranjan Singh, *Tribalism and the Tragedy of the Commons: Land, Identity and Development: The Manipur Experience.* New Delhi: Akansha Publishing House (pp. 239-257).

Somi, Mashangva. 2010. 'Why Manipur District Council Act is 'dangerous' for hill tribes', in *Mizoram Express,* 27 May.

About the Author

P.G. Jangamlung Richard is the Chairperson of North-east India Tribal Research Association (NEITRA). He is serving as Research Associate in the Department of Political Science, NEHU Shilling-2013. His areas of interest are *Theory of Autonomy and Women Rights in Afghanistan*.

Section – IV

Political Issues of North-East: Challenges and Strategies

Chapter 12

Liability of the Construction of Tipaimukh Dam in Manipur

☆ *Lungthuiyang Riamei*

ABSTRACT

Dams can offer huge benefits in terms of electric power generation and irrigation. At the same time, dams may destroy wild life habitats, drain wetlands and most importantly can displace thousands of people. First commissioned in 1984, the proposed Tipaimukh dam on the river Barak is mainly to control flood and generate hydroelectric power. The dam area lies in sensitive ecologically and topographically fragile region. There has been controversy between the Governments of India and Bangladesh over water rights and this has been an issue in bilateral relations. The chapter examines how the controversial project will have effect on biodiversity of the region.

12.1 Introduction

The proposed Tipaimukh Hydro project is located in the south-western Churachanpur district of Manipur (24°1"N and 93° 1"E). The dam will be 390m long and 162.8m high, across the Barak River of 500m downstream of the confluence of the Tuivai (*Tui* = water and *vai* = to wander) and the Barak (known as *Ahu* for the Zeliangruang Nagas and *Tuiroung* for Hmar people) on the Manipur-Mizoram border (see the map). The ethnic Hmar and other kindred people call this confluence as *Roungle Vaisuo.* The dam was originally designed to contain flood waters in the lower Barak valley but hydro power generation was later incorporated into the project. The project will have an installation capacity of 1500 MW and a firm generation of 412 MW.[1]

Initially, the project was first being taken by North-east Electric Power Projects (NEEPCO). But now, it will develop with a joint venture between National Hydroelectric Power Corporation (69 percent), Shimla-based Satluj Jal Vidyut Nigam Limited (26 percent) and Manipur Government (5 percent).[2] The dam will permanently submerge an area of 275.50 square kilometers. Earlier the proposed project cost was estimated at Rs 1,087 crores but now it has been revised and cleared for Rs. 5,163.86 crores by the Public Investment Board and the Central Electricity Authority of India. This project will be one of the largest hydroelectric projects in North-eastern India. With the construction of the dam not only destructions of environment but folklores and legendary of the indigenous people will be perish. The Indian government has offered the Manipur state 10 per cent free electricity (*i.e.* 40 MW) from the project. Before the undertaking of the mega dam projects it is necessary to re-examine how far this Hydro project is meant for development. The Tipaimukh dam project has been further delayed because the Bangladesh government will undertake 14 surveys on water life and wealth of Barak river. The survey will complete within two years and then the construction of the dam will commenced.[3] The meeting of the sub-group took place in Dhaka on 2 February 2013 and has asked Delhi to provide more information and data on the water flow of the Barak River to assess the possible negative impacts of the planned dam on the common river on Bangladesh.[4]

Some of the Indigenous people in different parts of the world have already displaced or disappeared due to climate change and have become environmental refugees. The argument is that construction of big dams like Tipaimukh is not solution to mitigate the lives of the indigenous people.

12.2 Consequences of Constructing Dam

Hydropower has traditionally been considered environmentally friendly because it represents a clean and renewable energy source. The term renewable refers to the hydrologic cycle that circulates water back to our rivers, streams, and lakes each year. Production of electricity with hydropower does not pollute the air, contribute to acid rain or ozone depletion because of carbon dioxide emissions, or (like nuclear power) leave highly toxic waste that is difficult to dispose of.[5] But this will not be same in the case of the proposed Tipaimukh Dam project. We have seen how the Narmada dam was constructed without the consent of the indigenous people and thousands of people were displaced. Neither their voice were heard by the government. Now, similar Narmada dam is in the process in the North-east India.

12.3 Changes in the Eco-system

Specific ecosystem impacts caused by single hydroelectric project largely depend on the following variables: i) the size and flow rate of the river or tributary where the project is located, ii) the climate and habitat conditions that exist, iii) the type, size, design and operation of the project, and iv) whether cumulative impacts occur because the project is located upstream or downstream.[6] Seeing all these conditions, construction of Tipaimukh dam will drastically change the topography in the entire region.

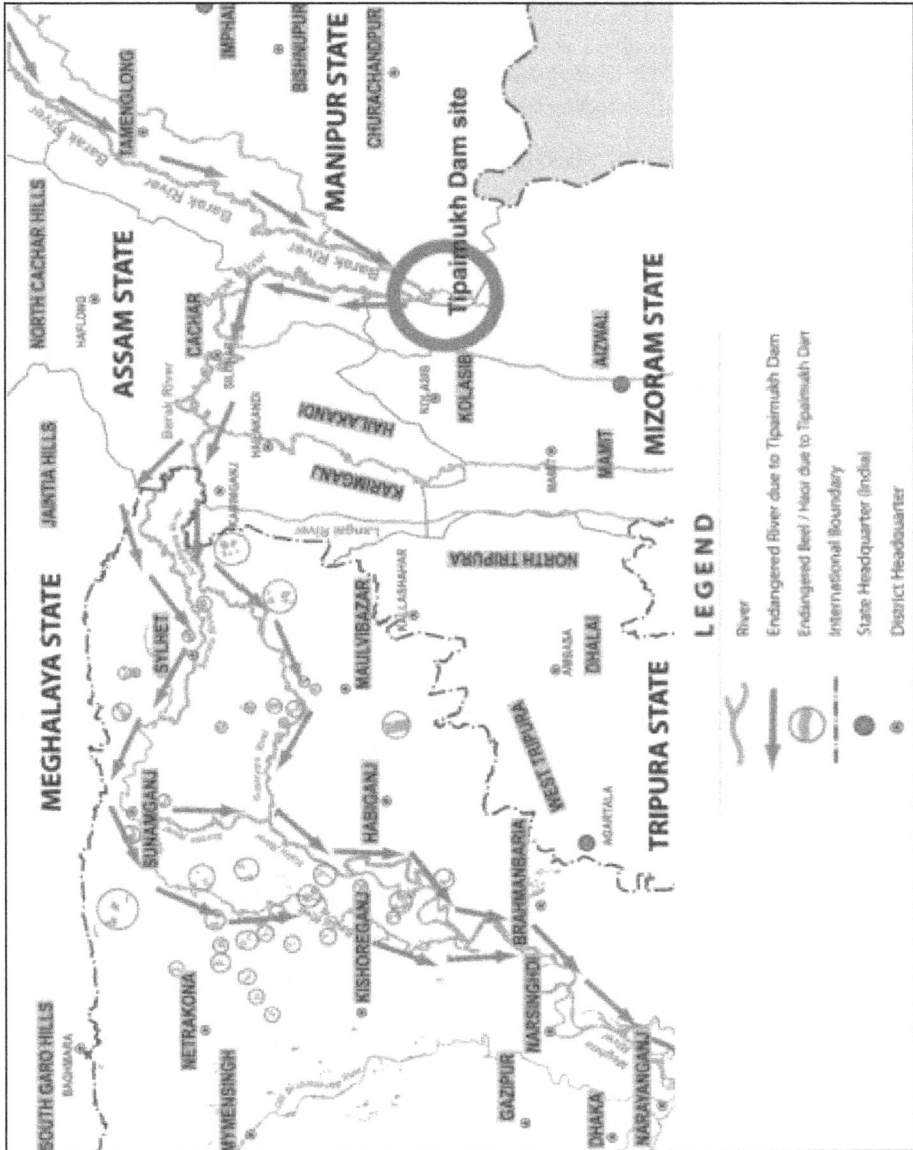

Source: Muddying the waters http://www.thedailystar.net/magazine/2009/07/02/cover.htm.

Hydroelectric projects do affect the ecosystems of rivers and their surrounding areas. The degree, however to which any project affects a river varies widely. Tipaimukh dam will have serious implications for the original inhabitants. The centuries old habitat in the region will be gone. Most of the displaced people will be the Hmar community (a kin to the Mizo people), and the Zeliangruang Naga.[7] These indigenous people live in the downstream and upstream of the Barak River which serves them as the vein. It is also the vital arteries for trade and transportation for the downstream inhabitants. It is feared that with the construction of this dam hundreds of fishing villages will be affected or submerged by the dam upstream.

12.4 Impacts on Cultural Heritage

The proposed Tipaimukh dam will bring losses and destruction, especially to people of Tipaimukh and Tamenglong living around the dam areas. Barak (*Ahu*) waterfalls and *Zeilad* Lake that are connected with the Zeliangruang history will be forever submerged underwater. All folklores and legends will have no living monumental proof for the coming generations. The dam would simply destroy their livelihood by affecting the entire agricultural lowlands on the banks of the three rivers namely; Barak (*Ahu*), Alang and Makhu which flow through Tamenglong and beyond. The township of Nungba sub-division of Tamenglong and the villages along the National Highway -53 (now NH-2) will be severely affected. Eight villages situated at the Barak valley will be completely submerged. More than 90 villages mostly in Tamenglong district will be adversely affected. About 27,242 hectares of cultivable land will be lost and roughly 50,000 people will be rendered landless.[8] The project once installed will submerge the exotic flora, fauna and the rich gene pool, since the area falls under one of the genetic hot spot zones of the world where rare biodiversity resources exist. The only virgin forest around Zeilad Lake where tiger reserve is located in the region will disappear.

Manipur, compared to its small size has already enough dams for the purpose of generating electricity and irrigation. But because of mismanagement many of the dams cannot produce what was envisaged for it. For instance, Loktak Hydro Electric Project could not supply enough power to the state due to poor management, but has destroyed thousands of hectares of paddy fields. Likewise, Khoupum Irrigation Project in Tamenglong district becomes a total failure resulting in the loss of thousands of hectares of wet paddy field and thousands of people displaced.[9] Seeing it from the two examples, Manipur which is considered as 'failed state', and also popularly known for 'corruption' is feared will bring any positive development by such mega project, rather the dam will bring more disaster for the local people.[10]

12.5 Impact for Bangladesh

Any mega-dam building will have impact on the downstream. According to some experts, if India builds the dam on the River Barak just 200 km away from the Bangladesh border, the water-flow of the River Meghna in Bangladesh will be hampered in a great deal. The Barak enters Bangladesh at a place called Amalsidh, near Zakigonj in Sylhet, after splitting into the Surma and Kushiara rivers. The Surma and Kushiara in turn unite near Bhairab to form the mighty Meghna. The Surma-

Kushiara-Meghna system supporting the livelihood of roughly 50 million people spanning 16 districts is one of the three main river systems in Bangladesh. Construction of the Tipaimukh dam will dry up 350 km long Shurma and 110 km long Kushiara River, thus choking the north-eastern region of Bangladesh.[11] As a result, the north-eastern Bangladesh will face severe environmental and economical consequences. It is considered that Sylhet, which is full of flora and fauna, will turn into a desert if this dam is built. Another thing to know is that there had been major earthquakes in the last 50 years on the location of the dam. But Indian government is going ahead with this project as they planned.

Since the deal was signed without any knowledge of Bangladesh, the Bangladeshi experts, opposition political parties and the media have blamed the government for failing to take diplomatic steps to stop the dam construction. They have also criticised the Sheikh Hasina government for its 'imprudence' of relying on India's "non-binding assurances" on the dam. And environmentalists have expressed grave concerns at the ecological, economic, and above all human consequences the dam would have for Bangladesh. Environmentalists and agriculture experts have warned that the dam at Tipaimukh across the cross-border Barak River would dry up rivers and water bodies downstream, rendering vast farmland arid, hitting agriculture and threatening food security in the north-eastern districts of Bangladesh.[12]

Navigation in river channels in the Meghna will be affected due to depletion of water flow and consequent sedimentation and severity of flooding during the monsoon season. Surface irrigation will also be in danger. The Meghna-Padma river will have lower flow, accentuating saline backwater intrusion in the Padma channel. Professor Anu Muhammad said that a total agricultural sector of around 20 districts directly and indirectly will be affected. The Barak-Surma- Kushiara-Meghna river system stretches about 946km. Around 669 km of this is in the Bangladesh portion. If India withdraws water, the fate of this whole river system will be threatened.[13]

After the project of Farakka Dam, India is starting another similar project which is equally dangerous for the ecology and Bangladesh. According to international laws, without the permission of the downstream river nation, one country cannot control the flow of a multination river. But it is very sad that India cares little about what Bangladesh has to say about this issue. The Indian government quietly floated international tender and completed preparations for the construction even without consulting Bangladesh.[14] Thousands of people and many organisations from both the countries are protesting against this project as it posts great threat to the ecology.

China unveiled a giant plan to relocate a vast population to make way for the south-north water diversion project in Yarlong Tsangpo river (Brahmaputra). It is this project which has caused great concern for India that it will affect the flow of Brahmaputra in the North-east India. The proposed dam is to be constructed at a place called Namcha Barwa on the eastern plateau of Tibet. The project will cause serious harm to the Brahmaputra River and adversely affect the low-lying areas in Bangladesh. India is proactive in addressing concerns with the Chinese government on the proposed dam over Tsangpo river, relaying its concern to Beijing in 2006.[15] The government of Bangladesh needs to be more proactive to the whole scheme of dam in the Brahmaputra river which will worsen water crisis in the country. Bangladesh faces a big challenge to confront the exploitation potentials of both China and India over the use of transboundary waters. There is primary concern for Bangladesh and the people of North-east India to explore all means to ensure China and India adopt multilateral, multiparty decisions over transboundary water use with due and full respect of rights and participation of indigenous peoples depending on waters. All States should refrain from unilateral and contradictory decisions over tansboundary waters disregarding downstream concerns and rights of the indigenous people.[16] The Chinese government also needs to consult neighbouring countries before going ahead for the multipurpose project.

If India has serious concern about building dam/power generation projects by the Chinese government in the Yarlong Tsangpo (Brahmaputra) River, then the Indian government should also re-think again before constructing Tipaimukh Dam. Instead of honouring the wishes of the people, the project authority and the state government have taken many undemocratic measures including the use of force (CRPF, The Assam Rifles, a State paramilitary wing deployed in North-east India for counter insurgency operations under draconian law- 'Armed Forces Special Power Act' AFSPA, now has extensively plan to provide security for the Project) that resulted in more rights violations and militarisation of the area.

A study carried out by the Helsinki University of Technology (Finland) on the Mekong River found that hydroelectric dams in China drastically reduced the load of fertile sediment carried by the monsoon floods, affecting the fertility of the Mekong's fisheries in Thailand, Laos and Cambodia. Experts suggest that the Tipaimukh dam may help control flash floods, but have warned of worse flooding to come.[17] The dam may control some of the annual flooding. But when there is a really big rise in water levels, the gates of the dam will have to be opened to save the dam itself. Then that will lead to a much bigger flood downstream.

12.6 Seismically Active Zone

The North-East region of India is one of the major seismically active zones of the world that includes Japan, Mexico, Indonesia, California, Taiwan and Turkey. Over the past decade and half, the issue of Tipaimukh dam has created a lot of disenchantment in regard to scientific, technical, environmental feasibility and socio-economic impacts of the dam. Tipaimukh and its adjoining areas are basically made up of Surma group of rocks. The rocks of Surma are mainly light grey to brownish grey generally medium to coarse grained sandstone having occasional shale and silt/ sand intervening bands between massive to thickly bedded sandstones. Conglomerate (loosely cemented pebbles and gravel) horizon at the base of Bhuban formation, though, can be.[18] North-eastern region falls among six major seismically active zones of the world tectonic setting. In this region, two typical tectonic settings are found resulting from the convergence between Indian and Eurasian plates. The trend of earthquakes shows that the regions which have experienced earthquakes in the past are more prone to it; the magnitude of future earthquakes may be uniform to the past ones; and the earthquake occurrence, geological data and tectonic history all have close correlation. The Tipaimukh dam site has been chosen at the highest risk seismically hazardous zone.[19] Such a consideration would reveal the nature and extent of the geotectonic risk being taken by constructing a mega-dam at Tipaimukh. The dam will not only cause problem for the people in North-eastern part of India but will also affect the neighbouring country.

Initially, the Government of Manipur had opposed the construction of the dam on the ground of seismological and environmental consideration. But the state legislative assembly unanimously resolved to rescind its earlier resolution to oppose construction of the dam on 15 December, 1999 and finally allowed the NEEPCO to go ahead for the project.[20] Meanwhile, the Central Electricity Authority of India has also given the final Techno-Economic Clearance to NEEPCO to start the project since July 2003. Due to incompetence of NEEPCO, the project has taken over by the NPHC and Satluj Jal Vidyut Nigam in 2010. In fact almost 90 percent of the Manipur's legislatures who agreed to sign Memorandum of Understanding for the construction of the dam didn't seek wider consultation from the affected people. One should also question how many indigenous legislatures are there in the Manipur's cabinet. Interestingly, some of the legislatures didn't know where the dam will be located.

12.7 Peoples' Response

Amidst mixed reactions from the local people and protest by some civil societies and non-state actors, the Government of Manipur and NEEPCO signed the Memorandum of Understanding (MOU) on 9 January, 2003. Debate and controversy centre on the magnitude of the dam and its possible impacts. Anti-dam activists argue that the dam is bound to uproot existing settlements, disrupt the culture and sources of peoples' livelihood and degrade environmental resources. On the other hand, pro-dams contend that the dam is expected to make an important and significant contribution to human development and its overall benefits in terms of power, irrigation, navigation, etc. The State authorities who are keen for the construction of the dam will never be aware of the land submerged, since their land will not be affected. Besides, there are some people (specially the contractors and politicians) looks only for their vested interest without much concern for the people.

Many non-state actors and civil societies have been campaigning against the construction of the dam. Organisation called COLNAR *i.e.*, Committee on Land & National Resources, comprising UNC (United Naga Council), NPMHR (Naga People's Movement for Human Rights), NWUM (Naga Women's Union, Manipur), ANSAM (All Naga Students' Association Manipur), The Sinlung Indigenous People Human Rights Organisation (SIPHRO), and Action Committee Against Tipaimukh Project, all these frontier organisations in collaboration with some Bangladesh NGOs including Sylhet based have raised issues on upper basin riparian and lower riparian affects.[21] They forecast the multifarious devastations on locals in loss of culture, habitats, flora and fauna.

The Affected People's Association of Tipaimukh Dam (APATD) based in Mizoram resolved that it would fight to the end to make the Indian government stop the proposed construction of the dam as it would displace a large number of people. Lalpekliana the APATD General Secretary said that it would submerge huge area of land displacing thousands of people living in the villages of Manipur and Mizoram borders. He emphasized that the rich biodiversity in the area would be destroy and no amount of money could compensate the loss of land.[22]

The North-east Dialogue Forum, a joint forum of more than 50 civil society groups also vowed to oppose the government's decision to build the Tipaimukh dam in Manipur. The forum Convener, U Nabakishore said that the Centre and the Manipur government are moving ahead with their plan to construct the controversial dam. Nearly 26, 000 hecters of forest land in Manipur will be affected and 78 lakh trees would have to be felled. The dam would severely impact on the livelihood of the local people and would also lead to climate change. There would be an obvious alteration of the rivers flow.[23] To continue their struggle, some activists also formed the Citizen's Concern for Dams and Development (CCDD) in 1999, which has become a strong platform for over 45 organisations. They also raised the issue on the three battalions of Central Reserve Police Force (CRPF) along with Assam Rifles which they planned to deploy by the government for forceful completion of Tipaimukh Dam in guise of fighting insurgency. CRPF and other para-military forces were engaged in an 'Operation Green Hunt' against tribals in central India who are protecting their rights.[24]

The Citizens Concern for Dams and Development (CCDD) has expressed its concern with the formation of a Sub-Group between the Governments of India and Bangladesh for conducting joint surveys to facilitate an early construction of the dam. The sub-group has finalized a Terms of Reference (TOR) for the assessment of the Tipaimukh Dam on 28th August 2012 at New Delhi, where each country is envisaged for assessments in their respective sides.[25] The CCDD has stated that any decision on the Dam construction cannot be bilaterally decided only India and Bangladesh as the land, rivers, forests and all resources in Manipur belong to all the indigenous people. On 11 February 2012, several community organization of Manipur had called for revocation of the memorandum of understanding signed among the government of Manipur, the National Hydroelectric Power Corporation and the Satluj Jal Vidyut Nigam Limited.[26]

On 7 February 2013, Tipaimukh Dam Affected People's Association (TDAPA), also resolved to intensify their struggle and reached at a meeting held at Rovakot Churachandpur in Manipur. The TDAPA decided to launch a full scale campaign against the building of the dam in the near future as this mega dam will wipe out the indigenous people, historical-cultural resources and their land.[27] The Hmar Students Association (HSA), powerful student's body in Churachanpur district of Manipur has also threatened to stop the implementation of proposed dam. The student's activists have revealed that they would sacrifice their life if the dam is constructed in their village.[28]

Concerned civil societies, village elders and the keepers of indigenous knowledge in the Tamenglong district, Tipaimukh area and the people in Barak valley have consistently voiced their urgent concern that when the dam is built, the entire valley area will be permanently submerge radically altering the ecosystem and leading to change in local temperatures.[29] Till now, many innocent villagers are not even aware about the much hyped and controversial dam where their ancestral land will be submerged once it has completed. There has been no meaningful public consultation nor there is any environmental management that plans in formulating, implementing and monitoring environmental protection (during and after the project) nor even for any rehabilitation and resettlement.

12.8 Violation of International Convention

Any project to build mega Hydro dams should take into serious consideration of the all factors and which should be recommended by the World Commission on Dams. The United Nations Declaration on the Rights of Indigenous Peoples (UNDRIP) 2007 and the International Labor Organization, Convention stipulates that indigenous peoples shall have the right to decide their own priorities for the process of development as it affects their lives, beliefs, institutions and spiritual well-being and emphasised that they shall participate in the formulation, implementation and evaluation of plans and programmes for development that may affect them directly. Government of India does not adhere to such international norms and standards that is crucial for the proposed Tipaimukh dam, therefore suffers a drawback.

Under International Law, India has no right to divert unilaterally the flow of the trans-boundary Barak river by constructing the proposed dam. In accordance with the provision of the United Nations convention on the law of International watercourses 1997, India is under an obligation to provide data and information on the condition of the watercourse, exchange information and consult with Bangladesh, if necessary negotiate on the possible effect of the planned Tipaimukh Dam on the condition of the water of trans-boundary river Barak. Bangladesh has signed and ratified this Convention, and therefore can bring India before the International Court opposing projects like Tipaimukh to be built unilateral by India.[30] India has disregarded some major provisions of the 1997 UN Watercourse Convention on the Article 5(1) Equitable Utilization, (7) No Harm Principle, (9) Exchange of Information.[31]

Due to discrimination and marginalization, indigenous peoples have borne a disproportionately higher share of the negative impacts of the dam. Indigenous peoples' rights are often not addressed in national legal frameworks and therefore ignored by developers, with the result that indigenous communities are excluded from decision-making arenas. The ILO convention 169 is a comprehensive and legally binding convention that exclusively covers indigenous and tribal people's right. This Convention is the principal instrument used to uphold the collective rights of indigenous peoples. Consultation should be undertaken between the respective government and indigenous peoples within the country. Under such guidelines, indigenous peoples should be consulted about the legislative or administrative measures that may affect them directly and also allowing them to participate in the formulation, implementation and evaluation of plans and programmes for national and regional development that concern them.[32]

The UN Committee on Elimination of Racial Discrimination has urged the government of India not to construct the Tipaimukh Dam in its concluding observation of the Seventh session from 19 February to 9 March 2007, and in its special communications made on 15th August 2008. The forum further urged authorities concerned to follow free, fair and prior informed consent of the people under the ILO convention 107.[33]

Therefore, Tipaimukh Multipurpose project is still a premature plan as the project has not been introduced to the people for a wider consultation and consent in accordance to their traditional way of decision making process. It is also not in consideration of existing regional and international human rights and developmental standards, including the framework for decision making for Dams. The foundation stone laid by the then Power Minister of India Sushil Kumar Shinde in 2006 is a disrespect and insensitivity to the democratic call of the people of Indigenous people of the North-eastern State both in India and Bangladesh.[34] The proposed project is just another death trap for the indigenous people. New Delhi has asked Manipur government to supply electricity to the neighbouring country Myanmar despite shortage being faced by the state.[35] In this aspect, the state government is likely to take up the Tipaimukh hydro project soon.

12.8 Conclusion

The construction of Tipaimukh Dam will be a violation to the democratic form of Government's functioning since the indigenous voices are not taken and considered. It is a total disregard and life threat to the *Zeliangruang* Naga and *Hmar* ancient indigenous heritage. The Manipur government's keen on the construction of the dam reflects partiality. In the overall assessment, the dam has more negative developmental effects on social, economy and ecologically rather than bringing positive benefits to the people. It is seen more of a destructive project especially to the people and environment by huge areas submersion and relocation instead of being a blessing for those benefactors. The project will change its age-old knowledge, traditions and the whole geographical setting by disturbing food chains in the flora and fauna, etc. The bulk of the power generated from the project will be exported outside the State to cater the demands of the Industries and corporate houses where a small amount would be meant for the locals. The process towards the Tipaimukh Multipurpose Project has been a history of violations of all established national and international norms for development projects. Like violations of human rights, procedural and legal defects in gaining environmental clearance. Tipaimukh dam have become a high profile issue in Indo-Bangladesh relations and has a huge potential to be thorn in future bilateral relations. A comprehensive technical joint study from both the governments of India and Bangladesh needs to be taken before construction of the mega dam in the India's North-east. A democratic form of government is where the peoples' concern should be taken into consideration and not going against the wishes of the people. Hence, this project needs wider consultation and approval of those indigenous peoples that would be affected.

Reference and Notes

1.   Rashid, Barrister Harun Ur. 2013."Tipaimukh Dam: Facts and Current Position", URL: http://www.priyoaustralia.com.au/articles/179848-tipaimukh-dam-facts-current-position.html, Accessed on 15 April 2013.

2. "NHPC replaces NEEPCO to build Manipur's Tipamukh Dam", in *Business Standard*, Shillong, 15 July 2009, URL: http://www.business-standard.com/article/economy-policy/nhpc-replaces-neepco-to-build-manipur-s-tipaimukh-dam-109071500183_1.html, Accessed 20 April 2013.

3. "Tipaimukh dam to be delayed", in *The Telegraph*, Silchar, 3 February 2013, URL: http://www.telegraphindia.com/1130204/jsp/north-east/story_16518561.jsp#.UYCTdKKBmnZ, Accessed 20 April 2013.

4. Rashid, Barrister Harun ur. 2013. "Tipaimukh Dam: What is the current position?", in *Dhakacourier*, 14 March 2013.
URL: http://www.dhakacourier.com.bd/?p=10617, Accessed 21 April 2013.

5. "How a Hydroelectric Project can affect a River". URL: http://fwee.org/environment/how-a-hydroelectric-project-can-affect-a-river/, Accessed 21 April 2013.

6. "How a Hydroelectric Project Can Affect a River", URL: http://fwee.org/environment/how-a-hydroelectric-project-can-affect-a-river/how-a-hydro-project-affects-a-river-print/, Accessed 25 April.

7. Zeliangruang Nagas are cognate tribes Zeme,Liangmai and Ruangmei. They inhabited in the Indian state of Assam, Nagaland and Manipur.

8. Kamei, Namdingpou. 2006. "Controversial Hydro Electric (Multi-purpose Project, 15 December.
URL:http://www.epao.net/epSubPage Extractor.asp?src=news_section.opinions.Opinion_on_ Building_of_ Tipaimukh_Dam. Accessed 23 April 2013.

9. *Ibid.*

10. For detail see, Failed State? *Times of India,* 16 October 2011.
URL: http://articles.timesofindia.indiatimes.com/2011-10-16/special-report/30285863_1_manipur-s-ukhrul-manipuris-imphal-valley, Accessed 06 June 2013.

11. "Tipaimukh High Dam". 2013, in International Rivers".
URL: http://www.internationalrivers.org/resources/tipaimukh-high-dam-3499, Accessed 23 April 2013.

12. Habib, Haroon, 2011. "In Bangladesh, Tipaimukh dam pact sparks fresh row", *The Hindu*, Dhaka, 21 November 2011, URL: http://www.thehindu.com/news/international/in-bangladesh-tipaimukh-dam-pact-sparks-fresh-row/article2644442.ece, Accessed 24 April 2013.

13. Alamgir, Mohiuddin. 2009. "India's Tipaimukh dam: Another Farakka for Bangladesh in the offing", 12-18 June 2009.
URL: http://banglapraxis.wordpress.com/2009/06/12/indias-tipaimukh-dam-another-farakka-for-bangladesh-in-the-offing/, Accessed 24 April 2013.

14. Alamgir, Mohiuddin. 12-18 June 2009. "India's Tipaimukh dam: Another Farraka for Bangladesh in the offing?
URL: http://banglapraxis.wordpress.com/2009/06/12/indias-tipaimukh-dam-another-farakka-for-bangladesh-in-the-offing/,Accessed 24 April 2013.

15. See, Dikshit, Sandeep. 2013. "One river, two countries, too many dams", in *The Hindu,* 2 April 2013, URL: http://www.thehindu.com/opinion/op-ed/one-river-two-countries-too-many-dams/article4570590.ece, Accessed 25 April 2013.

16. Yumnam, Jiten. "Transboundary water conflicts and Tipaimukh Dam". URL: http://classic.kanglaonline.com/index.php?template=kshow&kid=1604& Idoc_Session=b225ef4feb4356af0dd0f79dbc05eaa8, Accessed 25 April 2013.

17. Al-Mahmood, Syed Zain. "The Tipaimukh Dam Debate", *Bangladesh Environment.* URL: http://www.bangladeshenvironment.com/index.php/cyclone/17-cyclone, Accessed 25 April 2013.

18. Hasan, Mahadi. "Tipaimukh dam is a geo-tectonic blunder of international blunder". URL: http://www.chintaa.com/index.php/blog/showAerticle/17, Accessed 26 April 2013.

19. Ibotombi, Soibam. 2009."Tipaimukh Dam: A geo-tectonic blunder of international dimensions,"*BanglaPraxis,* Research, Policy Analysis, Solidarity, URL: http:// banglapraxis.wordpress.com/2009/05/page/2/, Accessed 26 April 2013.

20. Hmar, Isaac L, (2005), "Let there be Dam in Tipaimukh for the development of the surrounding villages", 17February 2005, *e-pao news,* URL: http://e-pao.net/ epSubPageExtractor.asp?src=leisure.essays.Isaac_Hmar.Let_there_be_Dam_ in_Tipaimukh, Accessed 27 April 2013.

21. As Tipaimukh is located in the interior of Manipur some of the legislatures who have signed the agreement didn't have the knowledge of the proposed dam. Above that they have signed the MoU in the interest of the party workers only.

22. "APATD resolves to oppose Tipaimukh Dam". 2013. Aizawl, in *Zeenews.* URL: http://zeenews.india.com/news/north-east/apatd-resolves-to-oppose-tipaimukh-dam_841369.html, Accessed 27 April 2013.

23. "Activists against Tipaimukh dam", 27 March 2013, in *Zeenews,* Karimganj URL: http://zeenews.india.com/news/assam/activists-against-tipaimukh-dam_766440.html, Accessed 27 April 2013.

24. See, Sethi, Aman. 2010. "Green Hunt: the anatomy of an operation", in *The Hindu,* 6 February 2013, URL: http://www.thehindu.com/opinion/op-ed/green-hunt-the-anatomy-of-an-operation/article101706.ece, Accessed 27 April 2013.

25. "CCDD says no Tipaimukh Dam", in *Manipur Mail,* 29 August 2012. URL: http:/ /manipur-mail.com/ccdd-says-no-tipaimukh-dam/, Accessed 06 June 2013.

26. "Public consent needed for building Tipaimukh dam", in *The Statesman,* Imphal, 31 August 2012.
URL: http://thestatesman.net/index.php?option=com_content&view= article &id=421843&catid=36, Accessed 28 April 2013.

27. "Tipaimukh Dam Affected People's Association (TDAPA)", in *The Sangai Express,* Imphal, 8 February 2013. URL: http://www.thesangaiexpress.com/tseitm-23530-tdapa, Accessed 28 April 2013.

28. "Hmar Students Association vows to stop Tipaimukh dam at any cost", in *The Times of India*, Guwahati, 30 January 2012. URL: http://articles.timesofindia.indiatimes.com/2012-01-30/guwahati/31004931_1_hsa-activists-tipaimukh-dam-project-site, Accessed 28 April 2013.

29. Villagers in Bamgaizeng and Venchiangphai in Tamenglong district have voice their concerned as their village will be submerge once the dam is constructed.

30. Khan, MA Muid (2012), "Waging legal battle against Tipaimukh Dam", in *The Financial Express*. Dhaka, 9 January 2012.
URL: http://www.thefinancialexpress-bd.com/more.php?date=2012-01-09&news_id=93125, Accessed 28 April 2013.

31. McCaffrey,Stephen. 1997. " Convention on the Law of the Non-Navigational Uses of International Watercourses", New York, 21 May 1997, *Audiovisual Library of International Law*. URL: http://untreaty.un.org/cod/avl/ha/clnuiw/clnuiw.html, Accessed 29 April 2013.

32. "Power to the People? Hydropower, Indigenous peoples' rights and popular resistance in Guatemala". *The Association for International Water Studies*. P.7-8. http://www.fivas.org/fivas/vedlegg/Power-to-the-people_FIVAS_2010.pdf

33. "United Nations: Tipaimukh Dam Must Stopped". URL: http://www.change.org/petitions/united-nations-tipaimukh-dam-must-be-stopped, Accessed 30 April 2013.

34. Yumnam, Op. cit.,

35. "New Delhi asks Manipur to supply electricity to Myanmar", Imphal, 12 April 2013, URL: http://e-pao.net/epSubPageExtractor.asp?src=news_ section.News_ Links.News_ Links_2013.New_Delhi_asks_Manipur_ to_supply_ electricity_ to_Myanmar_20130413, Accessed 30 April 2013.

About the Author

He is a Research Associate at the Centre for Air Power Studies, New Delhi.

Chapter 13

Politics of Collective Identity: Revisiting the Identity Puzzle among the *Zo* People of North-East India

☆ *Roluahpuia*

ABSTRACT

The contemporary politics of the North-east is shaped by ethnic politics which often go along with political struggles. A quick glance at the ethnic formation reveals that groping of identity by various tribes, sub-tribes and even clans who believe to have the same history, origin, migration, etc. are on the rise. This very process has resulted into both ethnic consolidation and dissociation which indeed is an ongoing process in the region. This has now shaped the entire discourse of the region and manifest in all realms including social, economic and political. This chapter tries to demonstrate the process by taking the case of the Zo people who are today sharply divided on nomenclature lines despite their struggle for social as well as political re-unification. The politics around nomenclature and the failure to agree on a single nomenclature has weakened the very Zo integration project. The paper therefore tries to delineate ethnic dynamics in the case of North-east India by looking the experience of the Zo people.

Keywords: *North-east, Ethnic, Tribe, Nomenclature.*

13.1 Introduction

Ethnic identities and ethnic formation continues to be an ongoing process in North-east India. The ethnic as well as linguistic plurality adds up to the contemporary politics which are played around the issues of identity. The experience of North-east India in relation to ethnic identity formation is a unique one and has its own peculiar feature which equally adds to the dynamic and complex processes of ethnic formation. This can be seen from the continuing consolidation and dissociation of identities from different perspectives ranging from tribe to religious lines. There is a process of both fusion and fission of identity. Over the years, it has shaped the entire discourse of the region and manifests itself in all realms including social, economic and political. Nonetheless, it is still a baffling question as to whether the region will ever see an end to this process.

A quick glance at the ethnic formation reveals that grouping of identity by various tribes and communities who believe to have the same history, origin, migration, etc. are on the rise. Although the lumping of different groups of people under one ethnic name is a product of colonial rule. It is also observed that there are communities which came together to form a larger identity by embracing a particular ethnic name even in post-colonial time. Taking the case of Manipur as an example which this paper is going to deal with, ethnic category such as Komrem[1], Zeliangrong[2] are formed by group of tribes and communities. In the case of the Zo people, the experience is slightly different again. The identity formation among the Zo people has a colonial imprint and thus till today, we find a tussle among them on nomenclature lines. They were known differently according to geographical settlements such as 'Kuki' in Manipur, 'Lushai' in Mizoram and 'Chin' in Chin Hills of Myanmar.

The need for such categorization was necessitated by a concern to subsume the enormous diversity into near and meaningful categories for both classificatory purposes and administrative convenience (Xaxa 2008:2). This identification into different names also leads to the outcome of putting them under different administration. Till today, they continue to be divided not only on nomenclature lines, but are also separated by national and international borders.

In line with this, this paper attempt to lay out ethnic identity formation in North-east India by specifically looking at the case of the Zo people. The politics of naming which was inherited from the colonial rulers and the administrative blunder of putting them under different countries have led to an identity crisis among the Zo people. The colonial imposed names such as the Chin-Kuki and the emergence of other generic names such as Khul, Eimi, Mizo, etc. has invited more problems than solving. The third section of the paper deals with these politics of collective identity by looking into the claims and counter-claims made by the protagonist of each of the specific nomenclature. On the other hand, a common belief which is undeniable is the fact that they belong to the same ethnic stock, having common history and origin which led to the rise of unification movements which shaped it in the form of Zo Re-unification movements. Although the movement failed to achieve much of its objectives till date, but it has brought the feeling of oneness and togetherness among the people. The last section deals with the trajectory of the unification movement tracing it from colonial period till the present.

13.2 Tribal Identities in Manipur: Three Cognate Generic Names

Manipur is a home of numerous communities both tribal as well as non-tribal. Likewise, in other parts of the North-east region, colonial authorities used the concept of ethnic identities to categorize the communities. The peculiar feature of identification of Manipur is that they used certain nomenclature to identify the communities which they thought are having close socio-cultural affinity and linguistic relations. In other words, the colonial administrators lumped together vastly different groups of people for administrative convenience where socio-cultural heterogeneity is the norm. Some of them include Naga, Kuki and Chin. Although these are entirely a colonial construct, the post-independent scenario has seen a turn in identity formation or construction. Thus, we have new collective identities emerging up such as the Zeliangrong, Zomi, Komrem. An interesting aspect which remains is the fact that the tribes which assert these new identities are part of the nomenclature under Naga and Kuki ethnic groups.

Over the years, particularly after post-independent era, identity formation and its assertion are on the rise. It had been observed that the tribes of Manipur have grown more conscious of their separate identity in recent times than they were before (Das 1985:15). The tribes of the state, small or large in number are all involved in the process. Both the processes of fusion or fission are underway which makes the outcome for one or the other more likely. Thus, amalgamation of different tribes to form a larger identity grounded on the principles or belief of shared history and other socio-cultural affinities is one aspect while the other aspect is that some identities are formed as a counter or resistance to the older ones. In course of these adjustmental processes some groups alienated from the bigger ones, while others found themselves in aligned to compose a larger social identity or aggregate (Gangte 2010:3). This chapter discusses both these processes taking the case of the Zo people who at present are not only identified and separated by different nomenclature, but even by political boundaries, both national and international.

13.2.i Kuki

The term 'Kuki' is a generic classification of ethnic group that spread throughout the North-eastern region of India, Northwest Burma and Chittagong Hill Tracts in Bangladesh (Haokip 2011:28). The term is used by the colonialist to identify certain groups of people living in the present hilly areas adjoining India, Burma and Bangladesh. It is accepted that the term 'Kuki' does not have any indigenous component nor it was known or used by the people to identify themselves. As Soppit notes (1887) "The designation "Kuki" is unknown to the Tribes now so-called by the plain people. The designation "Kuki" is never used by the tribes themselves though many of them answer to it when addressed, from knowing it to be the Bengali or plain term for their people". It was a term used by the Bengalis to describe them which literally means 'hill-men' or 'mountaineers'. To quote from Grierson (1904), he said "Kuki is an Assamese or Bengali term, applied to various hill tribes, such as the Lusheis, Rangkhols, Thadous, etc".

The term *Kuki* appears to have its first usage in Sylhet which in the present is a part of Bangladesh. The term first appeared in Bengal Rawlins writing of "Cuci's, or

Mountaineers of Tipra (cited in Ning 2009). The British applied the term Kuki as a common nomenclature for all the ethnic clans they came in contact within the region (Haokip 1998: 58). It was spelled differently by different authors in many different ways such as 'Coocies', Cuci's, 'Kookies' even though the name and term is alien to the people to whom it is applied upon-with different spellings including 'Kukis', the name perpetuated by British administrators such as Lt.-Colonel Shakespeare (Shakespeare 1912) and C.A. Soppit (1893) to indicate the migrants into Manipur State, Naga Hills, and the North Cachar Hills of India (VanBik).

An interesting part of the identification of these groups is the classification into Old Kuki and New Kuki. This was known to be introduced by Stewart who applied the categorization, *i.e.* Old Kuki and New Kuki. These Old and New Kuki is categorized on the basis of their arrival in Manipur. Some Old Kukis emigrated from the jungles of 'Tipperah', the hilly country south of Cachar, while others came from the adjoining areas of Burma (Das 1985:5). The tribes which are put under Old Kuki includes; the Aimol, Anal, Chiru, Chothe, Gangte, Koirao, Koireng, Kom, Paite, Purum, Simte, Vaiphei, Zou, Hmar, Biete and Hrankhol. On the other hand, the New Kukis refers to the latter immigrant group known as Thadou/ Khongjai which again is composed by numerous clans and sub-clans. These classifications on other hand reflect the rational for the use of these terms which is none other than administrative conveniences.

13.2.ii Zomi

Before proceeding, it is pertinent to explain the combination of two different nomenclatures. The tribes who advocated the use of the term 'Zomi' are the one who are discontent with the names given by outsiders such as Kuki and Chin. Das (1985) rightly observed this phenomenon saying that "In Manipur, we have examples of groups giving up old identities and accepting new ones". This is true in the case of Zomi. Yet, Zomi has not gained much significance till today as some groups of people, within the Kuki and Chin continue to identify themselves as Kuki and Chin. So, we often come across the word Chin in some literature or on the identity problems concerning these groups of people in the form of Chin-Kuki-Mizo, or sometimes put in short form as CHIKIM. However, since this paper concerns with the identity politics that surrounds the present day Manipur along with Mizoram, will purportedly use the term 'Zomi' in place of Chin because it has more local relevance in the context of analysis.

There are different explanations which led to the rise and popularity of the term 'Zomi'. Although, the term is a post-colonial product, it has gained significant popularity in the 90s onwards when ethnic identity and political wars are on its zenith in the state of Manipur. Going by the origin and rationale or explanations given on the roots of the term, it is believed to be first used in the Chin Hills of Burma with the adoption of Zomi Baptist Convention in 1953. An excerpt from the website of Zomi Re-unification Organization (ZRO) shows this, "The general meeting held on March 5-7, 1953 at Saikah village (now Thantlang township of Hakha area) was attended by 3,000 Christians. Even there by far the vast majority of delegates were from the Hakha area and there was not a single voice of support from LAIMI, but the ZOMI BAPTIST CONVENTION was born, named and based on the foundation of historical truth, confirmed by the General Meeting at Saikah with the most remarkable spirit of Christianity and unity never experienced before or since". The usage of the nomenclature "Zomi" itself attempted to debunk the earlier nomenclature "Chin" which the people find no relevance and having no indignity.

With this, Zomi is derived from the generic name 'Zo', the progenitor of the entire Zo race, and thus adopted the term "Zomi' (people of Zo). The protagonist of this nomenclature cemented their argument on the basis of the names by which their forefathers have identified themselves, which is Zo. Historical records have shown that the terms like Kuki and Chin were alien to them while they call themselves to be 'Zo', 'Yo, 'Jho', or 'Chou'. As Shakespeare (1912) notes 'The term Kuki, like Naga, Chin, Shendu and many others, is not recognized by the people to whom we apply it". However, the nomenclature Zomi has not gained much popularity among the Kuki or Mizo communities and continues to be limited in use as well as in terms of popularity.

13.2.iii Mizo

The generic name 'Mizo' is found to have its meaning as 'hill-men' or people living on high hills. The origin of the term Mizo has a slightly different interpretation and background. However, a well and great renowned historian from the Mizo community Dr. B. Lalthangliana believes that the term 'Mizo' is as old as the history of the people. To substantiate his proposition he quotes from another great historian saying that "K. Zawla put forward that our forefathers have called themselves Mizo while they were settling in Than-tlang (tlang=hills) in Chin State. This was believed to be between the periods of 1250-1400 A.D. Another great song writer Ms. Saikuti (1830-1921) in 1884 has used Mizo in one of her song. On the same line, the great Mizo Chief Bengkhuaia has named Mary Winchester as 'Zoluti' who was taken as a captive while raiding the tea garden at Alexandrapur in 1891. Lastly, Lalthangliana pointed out that the first book on Mizo dialect which came out in 1895 was known 'Mi-zo Zir Tir Bu' (Guidelines to the Mizo Dialect). His proposition objective was to support the argument that the term 'Mizo' is not the by-product of the post-independent Mizo movement.

Looking on the other side of the above proposition, it is said that the term 'Mizo' was used or popularized only after the Lushai Hills was official changed to Mizo Hills. As Gougin (2008) notes regarding the changed of names that 'Naturally, the word Lushai cannot be the nomenclature of the diverse tribes like Vaiphei, Gangte, Thado, Pawi, Lakher, Zou, Simte, Paite and Hmar etc. Hence, instead of calling Lushai they collectively choose Mizo as the alternative for the word Lushai'. Further he said 'the word Mizo was officially recognized by the Lushai sometime in 1946 synchronizing with the changing circumstances of politics'. On the same line, Burman (1992) observes that 'The tribes and sub-tribes preferred to identify themselves as Mizos and this was formalized when 'the Lushai Hills District (change of name) Act 1954 was passed by the Parliament. Under the Act the name of the erstwhile Lushai Hill District was changed to Mizo District with effect from 29th April 1954'. This change of name was done due to the changing nature of politics, and the term Lusei being limited to a particular tribe only. Therefore, Mizo was adopted by the people on the ground that it can cover the numerous other tribes, clans and sub-clans. Interestingly, in popular perception, the term is not exclusionist in the sense that it does not refer to any particular clan/group in a restrictive way (Changsan 1999: 233).

The nomenclature of Mizo has gained much popularity with the increasing popularity of the *Lusei Duhlian* language which is now adopted as the lingua franca of the Mizo. The adoption of *Lusei Duhlian* as the lingua franca was instrumental in shaping Mizo identity and continues to act as an identity marker. As Zou (2010) puts it 'language is the very core of modern Mizo identity. The emergence of *Duhlian* dialect as a standard language prefigured the formation of Mizo ethnic identity'. Also, its acceptance by the people as pointed out by Khamkhenthang (1986) a noted scholar of the Zo community remains pertinent to mention. Khamkhenthang said 'in the case of Mizo, it is a word born from inside with clear meaning in local dialects. It is not an imposed word. So it has firm rooting in the social fabric'.

From the above discussion, the issue of different nomenclature reflects the nature of identity politics as well as identity crisis among the Zo people. Though different nomenclature are being used and voiced, it must not be mistaken nor confused that the people, tribes or clans who are under these nomenclatures are people of the same origin. For example, it is an accepted belief that they originate from 'Chhinlung' or 'Sinlung' which is believed to be a cave located in South-east of China. Moreover, the close socio-cultural affinities and the linguistic relations they have make it undeniable that they are from the same family. Despite of this accepted belief, they could not come to any agreeable conclusion to which nomenclature should be used commonly for all.

13.3 The Politics of Collective Identity: Competing Identities and Contesting Claims

In simple the different nomenclatures of Kuki/ Mizo/ Zomi are formed by a conglomeration of similar groups. They in actual are shared identities, and to the different tribes which fall under these nomenclature, they are not exclusivist neither they have sharp boundaries. It is therefore common to see tribes, clans or sub-clans to fall in either categories or more than two in certain cases. This can be an outcome of numerous factors, geographical location, political or even at times economical. The various aspects by which these politics interplayed are at the local level and the factors that contribute to the acceptance and non-acceptance of a particular nomenclature will be analyzed from both historical as well as contemporary perspective.

As mentioned above, these nomenclatures are composed by agglomeration of different tribes. It is a known fact that these different nomenclature refer to the same people found to have similar history. They are closely interrelated tribes having many common characteristics which cannot separate them from one another (Vaiphei 2008:162). However, these different nomenclatures though frequently used interchangeably are having different meanings. As discussed in the previous section, the roots and period of the emergence of each of them vary in time and space. However, here an analysis is done on what ground one nomenclature is preferred while other is discarded, to help us understand the specific forms as to why there are ethnic hostilities arising out among them.

To those who embraced the nomenclature of Zomi, both Kuki and Mizo are 'mistaken identity'. For the term Kuki is seen to have no indigenous component rather an imposed name which is derogatory while the latter Mizo is seen to be grammatically incorrect. As stated by noted scholars 'The word Kuki was supposed to be given by the Bengalis. Perhaps the Bengalis found them culturally 'unsophisticated', because Kuki literally means 'wild hill people'. The word 'Kuki' was a derogatory term which was given by outsiders'. Similarly, the term 'Mizo' is found to have its limitations on various aspects. Gougin (2008) said that 'Mizo' literally means 'Man-hill' (*mi* means man, *Zo* means hill) which grammatically sounds incorrect. But the word 'Zomi' when literally translated means 'Hillman' or 'Highlander' which sounds right. He further notes 'Candidly speaking, there is no difference in its meaning but the difference is that of formation of the word grammatically'.

However, the term 'Zomi' itself is challenged on various grounds and is found to have its own limitation. One major limitation is that it does not have any official recognition. It is expected will take some time to be recognised by the majority. **Ro (2007)** observes that 'The Paite tribe of Manipur advocated the term 'Zomi' by linking the suffix '*mi*' meaning people to the word '*Zo*', but also failed to receive any recognition from others and ended up being recognized as the name for the people who advocated this view'. Moreover, the same tribe which embraces 'Zomi' is still put under Kuki even today (See Haokip 1998; Gangte 2010). Countering the claim of grammatically incorrect explanation the protagonist of the Mizo nomenclature argued that when translated can simply be put or understood as 'people of Zo' and so they do not find any error in using the term 'Mizo' for all cognate groups.

Within these nomenclatures, there are no boundary lines which delineate affiliation or sticking to any of the nomenclature. In other words, there are no particular inclusions or exclusion strategies for a tribe, clans or sub-clans to affiliate to any of the given nomenclature. With this, what is observed is the continual 'ethnic switching' of particular tribes from one nomenclature to another. This clearly reflects that these nomenclatures do not have fixed boundaries. The process of 'ethnic switching' has been more frequent in post-colonial India and is still going on. An interesting aspect of this switching process is that it did not take place within the three nomenclatures alone.[3] Further, this shift has resulted in a shift to political loyalty when all the tribes are found to have their own political movements. On the other hand, there are also tribes which began to assert their independent identity exclusive of any nomenclature mentioned.

In the State of Manipur particularly, tribes are categorically put under Kuki, Naga and Meitei ethnic groups. But over time some tribes like Anals, Aimols, Chiru, Moyon, Chothe, etc. which are previously identified under 'Kuki' prefer to be identified as Naga leaving aside tribes which embrace 'Zomi'. Kipgen (2008) with regards to these ethnic switching explained that "In recent years, some Kuki tribes have seemingly been assimilated into Naga fold partly due to geo-political advantages, and partly due to the Naga armed movement".

Another interesting case is the decision of the Gangte leaders who decided to join the Mizo fold.[4] This move was made on 9th February, 1999 onwards at Chiengkawnpang, Churachandpur District, Manipur when all the apex bodies of the tribes such as *Gangte People's Council* (GPC) along with its organs namely; the *Gangte Youth Association* and the *Gangte Students Organization* decided to merge with the *Mizo People Convention* (MPC), *Young Mizo Association* (YMA) and the *Mizo Zirlai Pawl* (MZP) respectively, although there were resentments among many cognatic groups. Jusho said 'the Kuki were not happy with the development and termed the merger pact as an act of inducting Gangte into Mizo nomenclature' (2004:41). As part of resentment P.S Haokip (2006) sent a letter to the *Mizo National Front* (MNF) alleging that 'the Young Mizo Association (YMA) accepted the Gangte group as a part of Mizo, thereby undermining Kuki unity'. This indicates that the smaller tribes are oscillating between these big nomenclatures. Moreover, it reflects that these nomenclatures are not a solution to the problem of identity imbroglio but rather aggravates the identity crisis among the Zo people.

Another important feature that marked these nomenclatures is the 'overlapping of identities'. There are tribes, clans and sub-clans like Hangshing, Guite, Singson, Ngaihte, Changsan, Hmar, etc. found present in either among the Kuki or Mizo or Zomi folds, due to geographical setting in most cases. In Mizoram most of the clans and tribes who are in Kuki and Zomi folds in Manipur are under the Mizo fold, although the situation is completely different in the case of Manipur. Thus, there is an overlapping of identity based on geography or territorial locations. Despite this, even Mizoram is not free from it although the numerous communities accept the term Mizo as their ethnic name. Burman (1990) cites the case of Hmar tribe of Mizoram that 'To the north lives the Hmars who claim to be autochthones; but Hmars who live in the central region identify themselves as Mizo'. Hmar is a recognize tribe in Manipur but a clan for Mizo's of Mizoram. This reflects the continual overlapping of identity among same group of people.

From the above discussion, there are two inter-related phenomenon; one is identity crisis and the other is the search for an appropriate identity. Both processes continued to shape the present ethno-political situation as the same dilemma implies for many tribes of the North-east too. On the other hand, currently these nomenclatures are gaining roots since most of these tribes are armed now in unifying the group, thereby intensifying the already ethnic division. Therefore, to accept the unification for one single nomenclature among these cognatic groups is rather bleak at least at the foreseeable future.

13.4 Zo-Reunification Movement- Beyond the Nomenclature Way

Amidst the division on nomenclatures, what is less known is the Zo-Reunification movement which has been going side by side till today? The movement not merely sought to bring the Zo people under one nomenclature, but demand for the integration of all Zo people.[5] The Zo integration project has colonial roots when they are put under different administration by the colonial rulers. In post-independent India, the relegation of them as ethnic minorities has further pushed them to continue with the struggle. What necessitates such movement to occur is the colonial wrongdoing of identifying differently with different names and putting them under different administration. Below is a brief outline on the Zo struggle.

The Zo re-unification process may be viewed from two broad perspectives. First is the search for a common and acceptable identity, and second is the armed and non-armed struggle. The nomenclature politics going on and the birth of new identities are the outcome of an attempt to unify into one nomenclature.[6] The movements launched by any group be it the Kuki or Mizo have the same objective.

Beginning with Kuki movement, various attempts are being made by the past leaders to unify under an accepted nomenclature. Of them, various nomenclatures like Khul, Eimi, CHIKIM have emerged over time. Though all of them failed to gain much popularity and significance, yet the derivation of the term 'CHIKIM' is very interesting. The term was coined by taking the words; 'Chi' from Chin, 'Ki' from Kuki and 'M' from Mizo which makes up CHIKIM, signifying the complete tribes or all cognate nations' (Gangte 2010:226).

Similarly, another proposed nomenclature is the term 'Khul' which existed from 1940s. The word "KHUL" means Cave, and "KHULMI" means Cavemen, thus "KHULMINAM" literally means the Nation of Cavemen. The adoption of 'Khul' was in direct retaliation against the term 'Kuki', and those tribes who embrace it established an organization known as *Khul Union*. As put by Kamkhenthang (1986) 'It (Khul Union) was a political organization in opposition to Kuki'. In its early days of adoption, it was able to garner a lot of appeal among the people.

However, such nomenclatures were not able to get much popularity and acceptance among the people like the nomenclatures of Kuki, Mizo, Chin and lately the Zomi have already gained stronghold among some communities. In other words, the nomenclature such as 'CHIKIM', 'KHULMI', or 'EIMI' failed to get mass appeal. However, the emergence of these different nomenclatures reveals the strong desire for unification among the Zo communities under one name. However, the weakness is that they did not campaign strongly to unite all under the name 'Zo' itself.

However, of recently the Zo people have demanded and struggled for the unification of different Zo communities to be under one administration. It initially began during the colonial period when the Chin Lushai Conference was held at Fort William, Calcutta on 29th January 1892. There, they expressed the desire to be put under one administration.[7] Convened by the British, the conference is credited as the first serious attempt of the Zo people to find ways in bringing about administrative and political unification to territories occupied by them and which are by now apportioned between the independent states of Bangladesh, India and Myanmar (Khamkhansuan 2009:272). But the major outcome of the Conference was the amalgamation of North and South Lushai Hills while the major demands for the unification of the Zo people continue to remain unfulfilled till date.[8] Piang (2012) a noted scholar among the Zo community states that 'Even after India and Burma attained independence from the British Empire nothing was done to implement the resolutions of the Chin-Lushai conference. This is how knowingly the sentiment of the ethnic communities and their cultural unity were undermined. As such, the problem of the Zo people remained unsolved, which resulted in the rise of Zo nationalism that culminated in various ethnic mobilizations'.

The first mobilization developed with the rise of *Mizo National Front* (MNF) movement. Although the movement has its roots in the famine that plagued the then Lushai Hills, however the movement later focus its main objective on the integration of the Zo inhabited areas. This is seen from the organization objectives 'To serve the highest sovereignty and to unite all the Mizo to live under one political boundary'. In pursuance of fulfilling their desired objectives, the MNF also submitted a memorandum in 13th October 1965 to the Prime Minister of India. The first line of the memorandum states 'This memorandum seeks to integrate the case of the Mizo people for freedom and independence, for the right of territorial unity and integrity and solidarity; and for the realization of which a fervent appeal is submitted to the Government of India'. The movement rekindled national sentiments throughout Zoland and many young men from all corners of Zoland joined the movement and fought for the Zo rights (Keivoim 2008:198). Although the armed struggle received a strong support from the masses, it failed to bring all the Zo inhabited areas under one administration with one stroke of a pen with the signing of Mizo Accord in 1986 between the then MNF and the Government of India (GoI). This was considered to be one of the greatest betrayals by the Hmar, Kuki who fought shoulder to shoulder with the Mizos' during the 20 long years of the movement.[9] Experiences, however bitter they may be, have shown that ethnic aspiration cannot be neglected and overlooked, as the cost is immeasurable (Lamkhanpiang 2012:4).

The movement for unification was brought back with the birth of Zo-Reunification Organization (ZORO) which was established in the year 1993 following the 1st World Zomi Reunification Convention at Champhai, Mizoram jointly organized by People Conference (PC) of Mizoram and Zomi National Congress (ZNC), Manipur. Though it went unnoticed by the Indian press, the convention became the forerunner of the Zo Reunification Organization with the avowed object of unifying the Indian state of Mizoram and the Chin state of Myanmar where the Zo descendants are predominant (Pau 2007:187). In the meeting, it was reaffirmed that they should strive for integration of the Zo communities by unanimously declaring as 'We, the people of Zo ethnic groups, inhabitants of the highland in The Chin Hills and Arakans of Burma, The Chittagong Hill Tracts of Bangladesh, The Mizoram state and adjoining hill areas of India, are descendants of one ancestor' (ibid.:187). The significant about the convention is that the Zo people on both sides of the international boundary met for the first time over an issue that they had long been aspiring separately. It was in this convention that Zo-Reunification Organization (ZORO) was born and which continued to work tirelessly for the unification of all the Zo people at all levels; social, political and economic.

Zo-Reunification Organization has put the re-unification of all the Zo people at its top agenda to regain the Zo National identity. Since its inception, the organization has been involved in various processes in achieving the long dreamt goals of the Zo people. Following its formation, the organization had submitted a memorandum to the United Nations on 20th May 1995, basically seeking support and intervention of the United Nation by providing historical facts and appealing the United Nations to correct historical injustice meted out against these communities. Further, the memorandum itself states 'Re-unification of the members of this family known today under various names and nomenclatures falls within the subject matter with which the Atlantic Charter and the objectives of the United Nations which are most concerned seriously'. The organization also continued to represent the Zo communities at the United Nation Permanent Forum on Indigenous Issues from 1999 onwards. The ZORO continue to work and stick to its principle of struggling and achieving its objectives through non-violent means.

Other organizations with the same objective have also emerged over the years like the *Kuki National Assembly* (KNA), *Paite National Council* (PNC) etc. These organizations have strived for the integration of Zo inhabited areas and have repeatedly submitted memorandums to the Government of India (GoI) for achieving its objective. In this way, various organizations both armed and non-armed have sprouted over the years attempting to re-unify the Zo people who are divided both by state (within countries) and international boundaries.

Thus, it is understandable that the Zo people despite divided by various nomenclature is fighting more or less for the same agenda. But the problem of living under different administration and the increasing identity politics among themselves has further put a challenge in achieving their objectives. Although the majority agree that they have the same origin, culture, tradition, etc. they failed to come together under one ethnic name or single nomenclature for reasons of different opinion. This has put hurdles to a large extent in their movements for unification contributing to the failure of the objective.

13.5 Conclusion

From the above discussion, it is seen that the increasing identity based movements witness today in the North-east has deep historical roots. Ethnic formation and identity based issues faced today in the region is shifting from time to time. The terms of discourse and the nuances through which identities are taking shape in the region is well exemplified by the experience Zo people. This spirit of unification is largely instilled among the Zo people by other ethnic groups like the Naga, Bodo, etc. who over the years have seen the projection of inter and intra based identity. Such identity formation reflects the process of fusion and fission of social formation, besides unraveling the current identity crisis among the Zomi communities.

The search for an appropriate identity among the Zo people flows in parallel with the political struggles. However, the political struggles are weakened itself by the failure to arrive at one nomenclature due to the attachment of personal sentiments. But this political struggle for unification strengthens again when different political movements fought on this nomenclature issues.[10] This implies that the ethnic distancing and distrust that had among the same group of people are brought together again in one hand while on the other hand, the boundaries of the group becomes more and more rigid due to this. This is now a common popular phenomenon not only among the Zo people, but largely among different ethnic communities in the North-east. Thus, under such circumstances, the idea of Zo re-unification remains bleak at least in the foreseeable future until and unless the people consentaneously give in the idea for it.

Endnotes

1.   Komrem is an ethnic label of five tribes- Aimol, Kom, Chiru, Koreng and Purum.

2.   Zeliangrong is formed by taking the ethnic name of three tribes, Zeme (Ze), Liagmei (Liang) and Rongmei (Rong) which altogether makes Zeliangrong.

3.   Tribes such as Anal, Moyon who earlier are identified under Kuki prefer to call themselves as Naga.

4.   Gangte tribe are officially put under Kuki group.

5.   The Zo people are put under three different countries viz. India, Burma, Bangladesh.

6.   The case can be taken of the formation of Zomi National Congress (ZNC) spearheaded by Gougin from the early 60s onwards.

7.   In the conference, a resolution was passed where the 'majority desire that the whole tract of the country known as Chin-Lushai Hills should be brought under one Administrative head as soon as possible. They also consider it is advisable that the new Administration should be subordinate to the Chief Commissioner of Assam.

8.   See Pau, P.K (2007) 'Administrative Rivalries on a Frontier: Problem of the Chin-Lushai Hills' that explained in details as to why the Zo inhabited areas were not put under one administration.

9.   See Kipgen (2006) on 'The Great Betrayal: Brief notes on Kuki insurgency movement' on www.kukiforum.com.

10. The contrast is that Zomi including its armed groups and civil society of Manipur demand for Autonomous Hill State while on the other hand, the Kuki are fighting for the formation of Kuki State.

References

Changsan, C. 1999. 'The Chin-Kuki-Mizo Ethnic Dilemma: Search for an Appropriate Identity' in Kailash, S. Aggarwal (ed.). *Dynamics of Identity and Intergroup Relations in North-East India*. Shimla: Institute of Advanced Study (pp. 230-244).

Das, R. K. 1985. *Manipur Tribal Scene: Studies in Society and Change*. New Delhi: M.C. Mittal Publications.

Gangte, T.S. 2010. *The Kukis of Manipur: A Historical Analysis*. New Delhi: Gyan Books.

Gougin, T. 2008. *The Origin of Zomi in Prism of the Zo People*. Lamka: Publication Board, Paper presented on 60th Zomi Nam ni Celebration Committee (pp. 363-366).

Haokip, P.S. 1998. *Zalengam-The Kuki Nation*. Zalengam: Kuki National Organization.

Haokip, S. 2011. 'The Kuki National Assembly: Historicity and Evolution', in *Alternative Perspectives.* VI(I), (pp. 1-9).

Kamkhenthang, H. 1986. 'Groping for Identity', in Dr. Hawlngam Haokip (ed). *In Search of Identity.* Imphal: Kuki Chin Baptish Union.

Kamkhenthang, H. 2011. 'Kuki Linguistic Groups in Historical Perspective', in Ngamkhohao Haokip and Michael Lunminthang (ed.). *Kuki Society: Past, Present and Future.* New Delhi: Maxford Books (pp. 1-16).

Khamkhansuan, H. 2009. 'Hill-Valley Divide as a Site of Conflict: Emerging Dialogic Space in Manipur', in Sanjib Baruah (ed.). *Beyond Counter-insurgency: Breaking the Impasse in North-east India.* New Delhi: Oxford University Press, (pp. 263-289).

Khamkhansuan, H. 2011. 'Rethinking tribe identities: The politics of recognition among the Zo in north-east India', in *Contributions to Indian Sociology.* Vol.45 (2), (pp. 157-187).

Jusho, P.T. H. 2004. *Politics of Ethnicity in North-east India: With Special Reference to Manipur.* New Delhi: Regency Publications.

Lalthangliana, B. 2011. Mizo Chanchinn Chik Taka Chhuina (A Critical Studies in Mizo History). Aizawl.

Lamkhanpiang, L. 2012. 'Ethnic mobilisation for decolnisation: Colonial legacy (The case of the Zo people in North-east India)', in *Asian Ethnicity.* (pp. 1-12).

Pau, Pum Khan. 2007. 'Administrative Rivalries on a Frontier: Problem of the Chin-Lushai Hills', in *Indian Historical Review.* Vol.4, (pp. 187-209).

Roy Burman, B.K. 1992. 'Crisis of Identity among the Mizos', in B. Chaudhuri (eds.). *Ethnopolitics and Identity Crisis.* New Delhi: M.C. Mittal Inter-India Publications. pp. 534-549.

Ro, S.C. 2007. Naming a People: British Frontier Management in Eastern Bengal and the Ethnic Categories of the Kuki-Chin: 1760-1860, (Dissertation, University of Hull).

Vumson. n.d. *Zo History: With an introduction to Zo culture, economy, religion and their status as an ethnic minority in India, Burma and Bangladesh.* Aizawl.

Xaxa, V. 2008. *State, Society and Tribals: Issues in Post-colonial India.* New Delhi: Pearson Publication.

About the Author

Roluahpuia has completed his M.Phil Degree from Tata Institute of Social Sciences (TISS), Mumbai – 400 088.

E-mail: roluahpuia90@gmail.com   Mobile: 09167752765

Chapter 14

India's Conflict Management Strategy in the North-East: The Case of Indo-Naga Conflict

☆ *Phyobenthung*

ABSTRACT

The policy of post-colonial Indian state towards the minority communities have been one of assimilation and forceful integration into the Indian Union. The conflict in the North-east is basically one of identity conflict. However, India seems to be lacking coherent and comprehensive policy when it comes to dealing with her minority nationalities. Though the Indian constitution provides certain safeguards to the minority communities, the inadequacy of these safeguards are well proven by the continuing struggle of the ethnic nationalities in the North-east India. Taking a case study, this article discusses the problem and prospect of India's strategy of conflict management in the Indo-Naga conflict. It argues that so long as New Delhi fails to address the basic 'nationality question', the strategy of containment will not bear any favourable outcome. Now that many militant groups have come to the negotiating table for peaceful resolution of their issues, it would be wise on the part of Government of India (GoI) to seriously consider negotiated settlement of the problems rather than pursing the policy of conflict management.

Keywords: *Geo-military, Insurgency, Dismissive, Negotiated Settlement, Co-option.*

14.1 Introduction: The Missed Opportunities

Are we going to miss the opportunity again? The journey of the Naga freedom struggle is simply tumultuous. If the goal of freedom struggle is one of achieving the desired outcome, then the Naga have never came so close to reaching that goal. We can easily pick three instances where GoI and Naga have missed out on the opportunities to settle the Indo-Naga issue. The first being, when the Nine-Point Agreement was signed in July 1947 between the GoI, represented by the Governor of Assam Sir Akbar Hydari, and the Naga National Council (NNC) leaders, represented by Naga tribes. Here, the differences in the interpretation by both sides over the last Agreement point led the Naga Nationalist leaders renounce the agreement and went ahead with outright declaration of Naga independence on 14th August 1947, a day before India's independence.[1] While the Naga assumed that they would be free to choose their destiny at the end of this ten years period (Iralu 2005:90), the Indian side assumed that Naga would be satisfied by the provision of the Sixth Scheduled of the Indian Constitution (Verghese 2004:88). According to Srikanth and Thomas the Indian side showed little regard for the aspiration of the Naga leaders and went ahead with the drafting of the Constitution of India disregarding the Nine-Point Agreement (2005: 5).

The second instance was in 1964, when the first ceasefire agreement was signed between the GoI and the NNC leaders. Under the initiative of the Naga Baptist Church Council, along with the involvement of three eminent and influential personalities - Jayaprakash Narayan (a Gandhian), Bimal Prasad Chaliha (Chief Minister of Assam), and Reverend Micheal Scot (a well known British Church leader) - it was a rare opportunity for the peaceful settlement of the Indo-Naga issue. But the entrenched positions[2] did not allow the impasse to be broken, and the ceasefire was over by 1972; though the Peace Mission got aborted earlier in 1966 with the resignation of Narayan and Chaliha and expulsion of Scot from India for allegedly taking a pro-Naga position (Chasie 2009:246). Observing this failure Rajagopalan commented that these talks proved fruitless, but the efforts of the Indian Intelligence Bureau to exploit Naga tribal rivalries did not (Rajagopalan 2008:15).

The third occasion is exemplified by the ongoing 1997 ceasefire and the peace talks processes in the last 15 years. Though the ceasefire and the negotiation are not abrogated yet, the outcome is more or less a foregone conclusion, as the strategy of the GoI seems to be that of conflict managing rather than resolution. During the "Consultation on the Indo-Naga Peace Process and Possible Outcome", jointly participated by both the parties, it is learnt that the onus for the solution of the Naga political problem lies with the Indian side as the negotiation is over, and the negotiators on both sides have already agreed on the contentious issues. However, the Congress led United Progressive Alliance (UPA) government at the Centre seems to be least interested in resolving the issue. It is learned that the Naga negotiators have already met and discussed with the cross section of the Indian political parties, and obtained assurance of resolution once the final settlement is tabled in the parliament; but the ruling government is mysteriously holding back for the reason best known to them.[3] The general opinion of the Naga people is that unless some dramatic changes happen at the Indian political setting, the fate of the current peace process is more or less sealed, thereby missing a golden opportunity yet again.

The question here is: are we going to keep playing the 'blame game' and miss another precious opportunity again? Is the GoI seriously considering a negotiated settlement? Does India have a coherent and comprehensive policy *viz-a-viz* her ethnic nationalities? What is India's policy towards her North-east ethnic minorities in general and Naga in particular?

14.2 India's Strategy: The Dismissive Attitude

Since, the beginning of the Naga struggle, the Indian state has tried her best to discredit the nationalist leadership. Even today this line of thinking is strongly contented on 'bringing the independentists to the mainstream', rather than going with an open mind to addressing each other's concern. This kind of understanding exists not only in the mindset of the successive Indian political leaders right from Nehru and Patel, but also within the Indian military and intelligentsia. Nehru, of course, in the later stage admitted, "I feel that we have not dealt with this question of the Nagas with wisdom in the past" (Nag 2009:53). Lamenting on India's approach towards the Naga political issue during the time of Morarji Desai, the article '*Exterminating Angels*' explained: Indeed, most Indian leaders do not seem to have learnt anything in the last quarter of a century of fighting insurgency in Nagaland (Economic and Politically Weekly 1977:1519-20). "If, as the Prime Minister says,[4] it were merely 'a few Nagas… giving trouble, harassing the villagers', then the insurgency would have been crushed long ago. The might of the Indian army and the air force need not have been engaged, unsuccessfully, to put down 'a few Nagas" (ibid.:1519). Dismissive attitude of the Indian PM was reflected throughout his speech in the Indian parliament. That, "a few Nagas" can be easily "exterminated", have unfortunately work reversed to the advantage of the independentists. It has been established since the 1960s by many including J.P. Narayan that:

"The Naga people are unquestionably a nation … while there can be no doubt that the Naga problem is not a law and order question, but a question of freedom for the Nagas, I have also tried to show that the Nagas freedom movement may take a different character if it is placed in the context of a union of self-governing states. And the kind of relationship between Nagaland and India may be negotiated" (Ibid.: 1519).

Had the GoI recognized this fact, the Naga problem could have been resolved long ago; hence, there would not have been an insurgency problem escalating in the North-east India, as the Naga insurgency is considered the "mother of all insurgencies" in North-east India.[5] Unfortunately, successive governments at the centre have chosen to ignore this fact and continue to pursue the policy of containment.

In the pursuit of this containment policy, the Indian middle class, whose voices are heard by the policymakers, are being fed by the writers with one-sided version of the Indo-Naga conflict. This thinking pervades in many of the Indian writings. We also find a good number of books written on the Naga by the Indian military officers, who have served as operational commanders in the region. These writings, no doubt, totally paint a distorted picture of the Naga. Their bias approach is, however, understandable, as serving in alien region inhabited by people, whose culture, tradition and custom are totally alien in the first place. These officers remained 'hostile elements' to the Naga throughout their entire service in the region. In fact for a long time, posting in this particular region is considered as a 'punishment posting'. So writings by such people filled with personal enmity at the first instance are quite justified. However, writings on certain people are in a way representation of these people, therefore it is totally unjustified to generalize and project such people as motivated just because of personal animosity. Such writings have flourished in India for quite a long time, the result of which, manufactured the mindset of the Indian middle class in their outlook and approach towards the region and Naga in particular.

Reacting to one such Indian writings, P. Pimomo (a Professor of English at Central Washington University) asserts on an article written by Kamarupee on Phizo is an unfortunate piece of Indian neo-colonialist asseveration (1990:2339).[6] However, it is only one among the tens and thousands (and certainly not the worst kind of Indian writing on Naga) that shows the hopelessly one-sided exercise of power in the production and dissemination of knowledge about Indo-Naga relations. It is true that many Indian scholars have also made eminent contributions in the field of post-colonial discourse and Subaltern studies, but "as it is, Indian writing on Nagas, with a couple of exceptions, falls into the category of work done by what Antonio Gramsci refers to as traditional intellectuals, or experts in legitimation" (Ibid.:2339). Thus, there exists a paucity of Indian scholars who objectively represent the Indo-Naga conflict. This factor largely contributes to more damage to the already compounded problems, as the Indo-centric representation is readily consumed by the mainstream generality.

14.3 The Strategy of Containment: Militarization and Suppression

In a seminar, Kr. Sanjay Singh argues that the 'Indian state loves the land but not the people' of the North-east India.[7] This assertion seems quite true. India's geo-military and strategic interests in the region have been established long ago: with Chinese invasion in 1962 and off-and-on claims of China in Arunachal Pradesh only invigorating India's insecurity in the region. Therefore, it is only rational that India will have a strategic interest in this region, particularly from security point of view alone. This has been recognized by every Indian policymaker right from Nehru, when he says: "... I consider it fantastic for that little corner between China, Burma, and India to be called an independent state".[8] However, India's strategy of looking at the region purely from geo-military angle has only alienated the people of this region.

New Delhi's response to the problems in the North-east began with a military suppression. Following the rejection of the 1947 NNC-Hydari Agreement by New Delhi, the NNC went on and conducted the Naga Plebiscite in 1951 to prove that Nagas are united and behind the demand for the independence. This was unacceptable to Nehru, who, on the one hand, was championing national right to self-determination at the international fora and, on the other hand, stubbornly adhered to the inviolability of India's territorial integrity. Naga issue was viewed as law and order problem, and dealt without any sensitivity to the concern or appreciation of history. Echoing Nehru's sentiments, the Assam government passed the Assam Maintenance of Public Order (Autonomous District) Act 1953, applicable to the Naga Hills District (Naga Hills District was then a part of Assam) to deal with the Naga independentists. Consequently all tribal councils and courts were also dismissed. This Act paved the way for the enactment of Armed Forces (Special Power) Act (AFSPA) on September 11, 1958, by the Parliament of India, initially for imposing in the state of Assam, and later extended to the other states of North-east and Jammu and Kashmir.

Unforgivable atrocities were committed by the Indian Army under the cover of the AFSPA. Under this Act rampant custodial torture and killings, mass rape, killing of innocent women and children, force disappearance, grouping etc. were committed by the Indian Armies. The most appalling provision of this Act is the impunity given to the Indian army, hence getting away from the eyes of justice. Several Commissions constituted by different groups, including the United Nations, have recommended for the abrogation of the AFSPA, as it is a draconian and repugnant law, and as it symbolises only the excessiveness of state power. The Indian government is however not very concern at the moment. Instead, as if AFSPA is not enough, the Nagaland Security Regulation Act 1962, National Security Act 1980, were also introduced in the State.

These military and national security acts have only alienated the people of the North-east and the effectiveness of these acts has shown the reverse. Despite fifty-five years of the enforcement of AFSPA, the insurgency groups have not only grown in numbers but in strength. Therefore in the final analysis, these acts have utterly failed the very purpose of eliminating insurgency in North-east India. New Delhi moreover does not seem to have an exit strategy from this quagmire.

The recent assertion by General (retd.) J.J. Singh that the "iron fist in velvet gloves policy" was implemented to curtail the so-called unlawful activities in the North-east is a bold admission coming from a retired army General, generally known for its constant mode of denial on the atrocities and human rights abuses by the Indian Armies. The former General was however of the view that "the insurgency problem in eastern Arunachal district and other parts of north-east (*sic*) could be resolved only after a permanent solution of the vexed Naga issue was reached".[9]

New Delhi's geo-military and strategic security obsession does not even spared the appointment of the heads of the democratic states in India. In the North-east the appointment of Governors are made purely from security perspective with an aim to monitor the functioning of democratically elected governments. Talking about the appointment of retired Generals as Governors in the North-east, Sanjib Baruah said: "Nearly all of them have either occupied high and sensitive positions in India's security establishment or have had close ties to it" (2005:66). For example, in Nagaland, in the past 20 years since 1993, out of the eight governors appointed in the state, one is retired Lieutenant-General and six are retired IPS officers who had headed the Intelligence Bureau or NSG at one point of their service time.

'The fact that so many of the appointees are men who have just shed their uniform and all the appointees have had fairly intimate connections with the security establishment cannot be mere coincidence. As appointees of the central government and as a facilitating agent in the counter-insurgency regime, such antecedents serve very practical ends, particularly in ensuring that the demands of security override the rules of democracy in the event of a conflict between the two' (ibid.: 68).

14.4 Preemptive Strategy: The Policy of Co-option, Downgrading and Assimilation

Analyzing the various accords between the GoI and the Naga, Kr. Sanjay Singh concludes that every subsequent agreements and accords systematically downgraded the demands of the Naga (ibid.:68). It is true if we look at all the previous accords. Initially, in the Naga-Hydari Accord of 1947, the newly independent Indian state recognized the right of the Naga to develop themselves according to the freely express wishes of the people.[10] However, this agreement, signed by the Governor of Assam representing the GoI, was unilaterally rejected by New Delhi, without review too.

Through the 16-Point Agreement of July 1960, Nagaland state was created within the Union of Indian but administered under the Ministry of External Affairs. However, it was brought under the Ministry of Home Affairs in 1972 without any reason, possibly due to the unilateral abrogation of ceasefire agreement with the NNC.[11] The Shillong Accord of November 1975, which was signed under military operational duress, did not pay any reward as the signatories who signed the Accord on their own individual volition were immediately sidelined and the insurgency movement gained further momentum. The Accord was also ineffective in as much as it was not officially approved and signed by the then FGN. It is considered by the Naga as a total sell out by the six signatories.[12]

In the current peace process too there are several instances of downgrading the Nagas by New Delhi. In 1997 when the ceasefire agreement was signed between NSCN (IM) and GoI, there were three conditions, namely: without any precondition; that the talk should be at the Prime Minister's level; and that the talk should be held in a neutral third country. Slowly all these conditions seem to have been compromised or ignored. The current talks are now held with a condition that any solution should come within the ambit of the Indian constitution; the Prime Ministerial level is now downgraded to Ministry of Home Affairs, GoI issue, and talks are now held randomly at New Delhi, Dimapur, Kohima, etc.[13]

How does one read such scenario? Skimming the history of India's post-colonial approach towards the Naga tells us that New Delhi has been persistently pursuing the policy of co-opting certain sections of the Naga society, thereby creating a rift among them. This policy of 'divide and rule' is as old as the Roman Empire, and the Indian state has perfected it from their British colonial masters. To a very great extent this policy was successful. The first instance of this co-option was in July 1960, when the demand of the Naga People's Convention (NPC) for the creation of state was entertained by Nehru himself (Das 2011:74). Indeed, the establishment of Nagaland state not only weakened the NNC but also continued to play crucial roles in the Indo-Naga conflict. With the establishment of Nagaland state, the Indian state claimed to have ascertained the wishes of the people. This is a significant achievement for India because New Delhi could afford to arrogate Naga issue as an internal matter in the international platform and justify her atrocities and human rights abuse.

The second instance of New Delhi's policy of co-option was in 1975. With the active initiative of the state government, some section of NNC signed the Shillong Accord, accepting the Indian Constitution unconditionally and even laid down arms. This policy however backfired as the disgruntled section of the NNC led by Th. Muivah and Isaac Chishi Swu along with Burmese-base SS Khaplang formed the *National Socialist Council of Nagalim* (NSCN) in 1980, which is by far, a more potent resistance movement than the NNC.

Again, in the current peace process, the conflict management strategy is more conspicuous. Whenever the negotiation seems close to conclusion, New Delhi finds one reason or the other to stall the negotiation or make strategic replacement of government interlocutors. At one point interlocutor Swaraj Kaushal openly accused Prime Minister Vajpayee of "thwarting his effort to bring the peace talks with the NSCN-IM to a successful conclusion" (See__ "Naga Peace Talk: Another Setback"). Kaushal thus acrimoniously accuses Vajpayee: "He has no perception of the problems of the north-east... you should never go back on your word. Unfortunately, this prime minister has no respect for his word".[14] Similarly, towards the year end of 2012, several newspapers reported of final settlement coming before 2012 Christmas. With the initiative of Joint Legislators Forum (JLF) of the Nagaland State Legislative Assembly and other Naga civil societies in 2012, there was speculation a final solution is in the offing.[15] However, with the replacement of P. Chidambaram in the Home Ministry with S.K. Shinde as the Union Minister, the rumors slowly subsided and vanished.

New Delhi, instead of seriously considering negotiated settlement, has been allegedly patronizing the different groups, bidding against the NSCN-IM. It has engaged the other two important Naga factions by entering into ceasefire although it has not entered into negotiation with any of these factions. The GoI is aware that the NSCN IM stands a better chance of bringing solution to the Naga problem, given that the Naga people, through the broad consultation involving all the tribal Hohos, Naga Hoho, civil societies and other stake holders, have given the mandate for carrying out the current negotiation to final settlement. Even the Nagaland Legislative Assembly has resolved to support the current peace process, and pave way for solution in case the current negotiation comes to its logical conclusion.

Speaking about the India's policy of co-opting certain sections of the Naga society, Dolly Kikon said "Governments at the centre have, by turn, adopted a policy of militarization and of extending grants to a small section of the local elite that it has co-opted in the task of governance" (2005:2833). From the very beginning the Indian state has adopted two-part strategy in Nagaland: counter insurgency operation and talks with the moderates simultaneously (Rajagopalan 2008:14). Divided into different tribes, it is relatively easier for the Indian state agencies to pick favourites among the Naga and appease them with economic and development benefits. Therefore, to some extent, this strategy of co-opting certain sections of people has worked amazingly well in the Naga society. However, the NSCN is also aware of New Delhi's game-plan. The NSCN manifesto states: "The pouring in of Indian capital in our country for political reasons has shattered the Naga people into a society of wild money, creating a parasitic, exploiting class of reactionary traitors, bureaucrats, a handful of rich men and the Indian vermin" (Baruah 2005:74). It is true that New Delhi's strategy of co-opting certain sections of the society is only creating corruption and destroying the fabric of the Naga society. As for now it seems to overlook the corruption that is happening in Nagaland.

It is true that the GoI, by a shrewd mixing of political and military pressures, has succeeded to a great extent in dividing, bribing and bullying the Naga people. Nonetheless the past settlements have also given rise to further complications since none of the settlements dared not touch the basic 'national question' that JP Narayan talked about. The Naga youngsters are now seemingly convinced about the futility of armed confrontation with India and many of them may be fading from the memories of the struggle for the right to self-determination. However, it should not be construed as a growing affection of India and her policies. The undercurrent believe is that Naga people's identity is different from that of India is argued by Satoshi that "their separate identity is even reflected in the consumption pattern of media products, which prefers to consume western media products instead of Hindi mass culture" (2011:53). According to him, Naga "have slipped out of the net bound by the slogan 'unity in diversity', despite the fact that the GoI is trying to incorporate North-eastern people into 'India" (ibid.:71).

14.5 Conclusion

What then is the way out from this conundrum? If one has to inaugurate a new era of peace that is comprehensive and final, one should address these salient points seriously:

i) The Issue of Self-Determination

There has been a change constantly in the interpretation on the notion of the right to self determination. Initially, it implies sovereign independent nation. Later on, territorial integration of Naga inhabited areas in India. This issue comprised an important point in the 30 Point demand of the NSCN-IM as well.[16] There is now a talk of Alternative Administrative Arrangement (AAA) for the Naga. Neiphiu Rio the Nagaland Chief Minister talks of it as "Emotional integration".[17]

The Naga independentists seem to have redefined their understanding of self-determination. The two nation theory is giving way for a unique federal arrangement with the Indian Union, an arrangement that gives some kind of 'Pan-Naga Mechanism' providing common socio-cultural and political identity. At this juncture it is important for the Indian state to find the kind of arrangement that the Naga aspire, instead of downgrading and bullying the independentists. Because New Delhi may succeed today in this policy, but ultimately unless the basic nationality question is addressed piecemeal solution would only create further problems.

ii) Repealing of all Anti-people Draconian Acts

As discussed earlier, draconian acts such as AFSPA have utterly failed to yield the intended result. Instead, as Chasie and Hazarika have pointed out "AFSPA has only pushed people away from government and the practice of democracy; and instead of reducing insurgency it has only helped to sustain it by violating people's human rights" (2009:27). The primary objective of the Indian army and paramilitary forces so far has been forceful assimilation of the Nagas into the Indian Union. This policy of assimilation is openly admitted by the Assam Rifles which declared that "the contribution of Assam Rifles towards assimilation of the North-east people into the national mainstream is truly monumental".[18]

Therefore, any settlement without taking into account of this aspect will only give a room for suspicion, and give a wrong signal to the people.

iii) The Development Agenda

Any form of solution to be final should address the economic backwardness of the Naga. The Nagas have missed the benefit of India's neo-liberal policy. Though Nagaland state has been enjoying the financial grants from New Delhi, these funds are inadequate and moreover it is not being utilized judiciously. It has been siphoned off by a handful of politicians and bureaucrats, and those people with necessary connections. Due to the non-settlement of the Indo-Naga issue, overgrounds and undergrounds are equally taking advantage of the situation. This situation is only creating a fertile space for corruption, widening the gap between the rich and the poor. A case in point is the demand of the people from *Eastern Nagaland* for separate state. Such movements may sometimes swing either ways; pro- or anti-India. Therefore, it is also in the interest of India to initiate active development process free of corruption and create a sustainable economy, which is accountable to the people. Now that India has realized the importance of connecting the Southeast Asian countries, a paradigm shift from geo-military to geo-economic priority should involve the North-east people so that the benefit of *India's Look East Policy* is shared by them.

In a nutshell, New Delhi must realize that the creation of the present state of Nagaland was never the ultimate goal of the Naga, although some skeptics may argue otherwise. Therefore, it is in the interest of both the GoI and the Naga to resolve the issue once and for all, which no doubt, demands a strong political will. Only then the Naga can also forge ahead along with other mainland people with confidence and without any sense of identity crisis.

Endnotes

1.   The Last Point, *i.e.* Point number nine talks about the Period of Agreement, that the Governor of Assam as the Agent of the Government of the Indian Union will have a special responsibility for a period of 10 years to ensure the observance of the agreement, at the end of this period the Naga Council will be asked whether they require the above agreement to be extended for a further period or a new agreement regarding the future of Naga people arrived at. See, "The Naga-Akbar Hydari Accord, 1947", (Viewed on 01-05-2013) http://www.satp.org/

satporgtp/countries/india/states/ Nagaland/documents/papers/ Nagaland_9point.htm

2. While the GoI insisted solution within the framework of Indian Constitution, the NNC leaders were not prepared to accept anything less than independence.

3. The "Consultation on Indo-Naga Peace Process and the Possible Outcome" was organized by Indian Civil Societies from 21-23 Jan. 2013 at the India Islamic Cultural Centre Delhi. During this time the Naga Hoho President stated that the negotiations are over; there is nothing to negotiate about as the two negotiators have already agreed on the contentious issues. The job now is for the Indian Parliament to decide. This stand is also reiterated by the other participants on both sides.

4. The author was referring to the Prime Minister of India Morarji Desai's statement in the Indian Parliament, of his meeting in London, the NNC leader Phizo on 16 June 1977. The Indian Prime Minister was criticized for using fascist's jargon in the Parliament when he spoke of "exterminating all rebels" of Nagaland. For detail see 'Exterminating Angels', in *Economic and Political Weekly*, August 27, 1977 (pp. 1519-20)

5. see Gautam Navlakha. 2003. 'Naga Peace Process: Larger Issue at Stake', in *Economic and Political Weekly*, February 22 (pp. 683-684)

6. In his article 'Passing of Phizo', Kamarupee allegedly made major distortions of Naga history and of Phizo's role in it. For more see, Pimomo, P. 1990. *Indian Writings on Nagas;* and Sajal Nag. 1994. *Mainstream Perspective on Nagas.*

7. A Seminar was organized by The North-east India Studies Programme, JNU, on 26th April, 2013, on the topic 'Accords as Instances of Domination: A Case Study of Naga Accords', presented by Kr. Sanjay Singh.

8. Nehru was referring to Nagaland in his speech in the Indian Parliament discussion after meeting the NNC delegation led by Phizo. Op.cit. Sajal Nag (p. 50).

9. General (Retd.) J.J. Singh was interacting with the media in Itanagar on the eve of his retirement as the Governor of Arunachal Pradesh. *The Morung Express*, May 24, 2013.

10. Bhaumik was of the opinion that, this attitude reflected the new Indian state hesitation to force the Nagas to integrate, a factor that changed after the successful integration of the princely states (2005:200).

11. The NNC and the FGN were declared as unlawful associations. (Viewed on 10-05-2013). http://www.angelfire.com/mi/Nagalim/chronology.html, .

12. The six signatories (I. Temjenba, S. Dahru, Veenyiyl Rhakhu, Z. Ramyo, M. Assa, and Kevi Yallay), on their own volition accepted the Indian Constitution unconditionally and promised to lay down arms. For more see, *Nagaland Accord: The Shillong Agreement of November 11, 1975.* (Viewed on 01-05-2013). http://www.satp.org/satporgtp/countries/india/states/Nagaland/ documents/ papers/ Nagaland_accord_the_shillong_nov_11_1975.htm.

13. For more see, 'Chronology of Peace talks with NSCN IM', (Viewed on 11-05-2013). http://www.satp.org/satporgtp/countries/india/states/Nagaland/data_sheets/Chronologytalknscnk.htm

14. For more see, _____(1999). 'Naga Peace Talk: Another Setback', in *Economic and Political Weekly,* July 31 (pp. 2120-2122).

15. *Nagaland Post* dated 2013, December 6 &15.

16. Though the NSCN-IM neither denies nor accepts this alleged demand of proposal, and no official White Paper was released; there were several newspaper reports on the 30 point demand. Some of the important demand includes, extension of Article 371 (A) to all Naga areas, some concession in foreign affairs relating to Nagas, Economic package and human resource development, provide assistance by India to the Nagas in Burma for basic infrastructure development, etc. The centre has also reportedly prepared 29-point counter proposal for discussion, which includes financial sops and greater autonomy. For more see, *Nagaland Post*, 04-08-2011, NSCN IM summits 30 PT Demands (04-03-2010) (Viewed on 25-05-2013) http://thohepou.wordpress.com/2010/03/04/nscn-i-m-submits-30-pt-demands/

17. Here Chief Minister Neiphiu Rio is certainly talking about some interim arrangement when he says 'though physical integration may be out of immediate reach, there was no reason why Nagas could not integrate emotionally'. *Nagaland Post*, 16-01-2009. Rakhi Chakrabarty was of the view that CM Rio's perhaps had in mind 'a body that unites the Nagas irrespective of which state they live', while referring to 'emotional integration'. *The Times of India* , 19-10-2012.

18. For more see, *The Assam Rifles: 178 Years of Glory and Sacrifice'* (Viewed on 16-05-2013) http://www.assamrifles.gov.in/history.aspx

References

Baruah, Sanjib. 2005. *Durable Disorder: Understanding the politics of North-east India*. New Delhi: Oxford University Press.

Bhaumik, Subir. 2005. 'The Accord that Never Was: Shillong Accords, 1975', in Das, Samir Kumar, (ed.) *Peace Processes and Peace Accords*. Delhi: Sage Publication. (pp. 203-221).

Chasie, Charles. 2013. *Nagaland*, (Viewed on 01-05-2013) http://www.ide.go.jp/English/Publish/Download/Jrp/pdf/133_9.pdf

Chasie, Charles and Hazarika, Sanjoy. 2009. 'The State Strikes Back: India and the Naga Insurgency', in *Policy Studies*. Vol. 52. Washington: East-West Centre.

Das, N.K. 2011. 'Naga Peace Parley: Sociological Reflections and a Plea for Pragmatism', in *Economic and Political Weekly*. June 18, XLVI(25), (pp. 71-77).

Iralu, Kaka D. 2005. 'The Fifty-Four Years of Indo-Naga Conflict: A Question of Internal Indian Ethnic Conflict or a Conflict between Two Nations', in Hussain, Monirul, (ed.) *Coming out of Violence: Essay on Ethnicity, Conflict Resolution and Peace Process in North-east India*. New Delhi: Regency Publication (pp. 87-113).

Kamarupee. 1990. 'Passing of Phizo', in *Economic and Political Weekly.* May 5-12, (pp. 983-984).

Kikon, Dolly. 2005. 'Engaging Naga Nationalism: Can Democracy Function in a Militarised Societies?' in *Economic and Political Weekly.* June (pp. 2833-2837).

Nag, Sajal. 2009. 'Nehru and the Nagas: Minority Nationalism and the Post-Colonial State', in *Economic and Political Weekly.* December XLIV(49), (pp. 48-55).

Nag, Sajal. 1994. 'Mainstream Perspective on Nagas', in *Economic and Political Weekly.* June 4, (pp. 1423-1424).

Ota, Satoshi. 2011. 'Ethnic Identity and Consumption of Popular Culture Among Young Naga People India' in *IJAPS,* Penerbit Universiti Sains Malaysia, September (Special Issue) Vol. 7(3), (pp. 53-74).

Pimomo, P. 1990. 'Indian Writings on Nagas', in *Economic and Political Weekly.* October 13, (pp. 2339-40).

Rajagopalan, Swarna. 2008. 'Peace Accords in North-east India: Journey Over Milestones', in *Policy Studies. Vol. 46.* Washington: East-West Centre.

Srikanth, H & Thomas, C.J. 2005. 'Naga Resistance Movement and the Peace Process in North-east India', in *Peace and Democracy in South Asia.* Vol.1(2) (pp. 57-87).

Verghese, B.G. 2004. *North-east Resurgent: Ethnicity, Insurgency, Governance, Development.* Delhi: Konark Publishers Pvt. Ltd.

About the Author

Phyobenthung is Assistant Professor in the Department of Political Science, Fazl Ali College, Nagaland. He is currently a Doctoral fellow at School of International Studies (SIS), Jawaharlal Nehru University, New Delhi.

E-mail: splotha@yahoo.com

Chapter 15

Nature and Patterns of Formations of Coalition Governments in Manipur 1990-2001: A Theoretical Perspective

☆ *Paolenthang Khongsai*

ABSTRACT

The chapter is an attempt to highlight the formation of various coalition governments in Manipur (1990-2000) in the light of two interrelated theories of coalition politics (office seeking or and policy seeking). Based on theoretical considerations, the nature and patterns of coalition formations in the state of Manipur are highlighted. Concentration on the nature and patterns of coalition formation is maintained to bring to light the nature and causes of stable or instability of the coalitions. Since the earmarked period is an era of instable coalitions, responsible factors are also dealt at length though not to the extent of exactness of the matter. Another field of investigation is the impact of change of guard at the Centre and its impact in state politics which often results to political instability. This area is pertinent to the study as most of the Indian states are seen dependent on the Central leadership for both political and societal advice. As the concluding part, some of the findings of the study are highlighted as observed by the author.

15.1 Introduction

Manipur is known for its ethnic and cultural diversity though small in size. The valley that constitutes roughly 10 per cent of the geographical area of the state is populated mainly by the majority of Meitei community, who constitute about 60 per cent of the state's population. The hill region, on the other hand, which is administratively divided into five districts, has a poly-ethnic population comprising of 30 recognized Scheduled Tribes and some other tribal communities who are still seeking official recognition of their names in the list of Schedule Tribes of India. Historically, the people of Manipur were known by the name of their separate communities throughout the 19th and the early part of 20th century. It was only after the advent of the British that the people residing in hills were divided into two groups under the generic names of Naga and Kuki.

Manipur was converted to Union Territory in 1963, and Statehood came much later in 1972. With the attainment of statehood, the strength of the Manipur Legislative Assembly was also raised to 60 member's seats.[1] The present Manipur Assembly has 60-members out of which 19 seats are reserved for Schedule Tribes and one seat reserved for the Schedule Castes. The state is represented in the Lok Sabha by two members, one member in the Rajya Sabha. Since 1972, Manipur had witnessed unprecedented political instability with a record number of 14 governments, 9 Chief Ministers and imposition of President's Rule for 7 times. No ministry had ever completed the normal full term of office till Assembly election held in 2004. Indeed, political instability had been a major feature since 1972 in Manipur. However, the major period of political instability had been witnessed during the period from 1990 to 2001. The period witnessed the formation of seven coalition governments which were led by both regional and national political parties, which was also been accompanied by imposition of President's rule on three occasions. Politics in Manipur, is thus seemingly characterized by the presence of a large number of political parties, frequent splits, defection, horse-trading and non-cohesive coalitions.

The chapter attempts to highlight the nature and patterns of formation of coalition governments in conjunction with political instability in Manipur although empirical studies can be undertaken from various perspectives. As widely conceived, the stability of a coalition government is deeply rooted in the nature and patterns of formation of the coalition itself. This is due to the fact that, in coalition, varying ideological combinations had been the hallmark of sharing the principal's office of the state. In this regard, two related theories of coalition formation namely; policy seeking theory and office seeking theory is used as hypothesis to examine the formation of seven coalition governments in Manipur. Further, an examination is also undertaken to understand the fallout of change of guard at the Centre and the consequences in state politics of Manipur. An examination of change of guard at the Centre and its fallout in state politics is highlighted where nepotism by the ruling parties cannot be ruled out in a complex political setup in federal countries like in India, especially when regionalization is on the periphery of politics, and that each ruling parties at the Centre had been seen to be wittingly desirous of establishing a stronghold even in the states.

15.2 Theoretical Considerations of Coalition Politics

The study of formation of coalition governments has attracted the attention of political scientists and the likes. They have tried to look into the formation of coalition from different perspectives and tried to predict which coalition would be more stable and durable. Analyses of government formation are also characterized by the motivational assumptions about the politicians. Politicians may be assumed to be motivated above all else by the desire to get into office - the office-seeking assumption. Or they may be assumed to be fundamentally concerned with the policies of government, wanting to get into office in order to influence policies.

Political scientists implicitly assume that the motivation of any politician can be described as either office seeking or policy seeking or some mixture of the two. The early models of government formation saw the making and breaking of governments as a competition over the allocation of the rewards of office like a fixed set of cabinet positions. In contrast, nearly all recent theoretical accounts are based on the assumption of policy seeking. At this juncture, it may be note-worthy that, since coalitions have played an integral part in the governmental processes of many western countries and European countries, the theories of coalition formations may be specific to those countries, and it would be highly fallacious to assume the universal principles of all government formation as homogeneous. It is therefore, necessary to distinguish between different types of political systems as differences do exist between them, and "any general theory that ignores them must be cast in abstract terms that would render it weak in prediction and superficial in explanation".[2]

Typically, a coalition had been regarded as a parliamentary or political grouping less permanent than a party or faction or an interest group.[3] In politics, it signifies a parliamentary or political grouping of different parties, interest groups, or factions formed for making and influencing policy decisions or securing power. In political system, coalition government implies the formation of an alliance or temporary union by some groups or political parties in order to exercise control over political power. Coalition governments are generally formed when no single party is able to command majority in the House. It is also generally accepted that a coalition can take place within the context of mixed motive in which both conflict and common interest are simultaneously present and must govern the course of action chosen. Coalition government implies the formation of an alliance or temporary union by some groups or political parties in order to exercise control over political power. Thus, a coalition is a phenomenon of a multi-party government where a number of minority parties join hands for running the government, which is otherwise not possible in a democracy based on majority party system. In such situations, various political parties join together to form a government without losing their separate identity. These political parties or groups agree to common minimum political, economic and social programmes, and when differences arise, any party or group is free to withdraw from the coalition. A coalition is thus, an alliance between two or more hitherto separate or even hostile groups or parties formed in order to carry on the government and share the principal offices of the state. Lane and Ersson (1996) classified theories of coalition as policy seeking model and office seeking model.[4] Office and policy motivation have been the dominant stimuli that induces coalescing behaviour among political parties.

15.3 Policy Seeking Theories

Policy-based theories considers ideological affinities as the key to the formation of coalitions and predicts minimum connected coalitions *i.e.*, coalitions that are composed of parties adjacent on the ideological scale, thus minimizing the ideological range, and within this limited condition minimum number of parties are needed for a majority. Ian Budge and Michael Laver (1986) assume the predominance of policy concerns over office seeking. They argued that "Parties may advocate policies either because they want to achieve office or because policies are what they are in politics to pursue".[5] Abram De Swaan (1970) a leading proponent also asserted that, considerations of policy are foremost in the minds of the actors.[6] The policy-based theorist considers ideological affinities as the key to the formation of coalitions.[7] Policy pursuit theory assumes that coalitions will be both connected as well as winning. This group of political scientist also argues that the benefits to the political parties, which are generated by a particular government comes primarily from the policies implemented by that government. Since government policies are public goods, the benefits that a party may receive from different governments are not related to the size of the governments, instead these benefits are related to the preferences of a party for the policies of one government compared with those of others. Each political party is considered to have a policy position identified at a particular point in the space representing the policy alternative which is most preferred by that party. Presumably, any party will prefer the policy that is closer to its own 'policy position' over the alternative that is further away.

15.4 Office Seeking Theories

On the other hand, office seeking theories basically held office *i.e.*, Ministerial berth or posts, as the primary goal of political parties and that all other objectives can be realized through the holding of office. Riker (1999) view that what parties fundamentally seek is to win, and in parliamentary democracies winning means controlling the executive branch.[8] This view of party behaviour was commonly assumed even in the early years. As Laver and Schofield puts it, "typically the yearning for office is seen as it was seen by Riker, as the desire to control some sort of fixed prize, a prize captured by the winning coalition and divided among its members".[9] Other political scientists associated with office seeking include Gamson, Martin and Stevenson, Von Neumann and Morgenstern etc, who have stressed office as an intrinsic concern for rewards such as power, prestige or a place in the limelight. The intrinsic rewards of office are seen as fixed. Thus, office-seeking model held office *i.e.* ministerial berths as the primary goal of political parties and that all other objectives can be realized through the holding of office.[10] Office seeking parties tries to maximize their control over political office of benefits. This is because of the fact that, portfolios afford parties the opportunity to influence those policy areas with which they are especially concerned. Budge and Laver also point out an ambiguity in the office seeking model. A party might stride to capture executive office because its leaders want to use to affect public policy, or because they think they can gain favour with the voters by exploiting the advantages of incumbency.[11] Office can have an intrinsic value, or it can have an instrumental, electoral, or policy value.

Today, neither pure office nor pure policy motivated explanation finds favour among students of coalition politics of the mix situations. The explanation that hold that office is used for achieving policy objectives, rather than policy being subordinate to office goals has been supported by many studies. In fact, this has become the dominated or hegemonic explanation within coalition studies today, but for scholars like Budge and Laver its office seeking and policy of a coalition.[12] According to the two scholars, office can be sought as an end in itself and as a means to fulfill policy objectives. Similarly, policy can be pursued both as an end in itself and as a means to achieve office. The two scholars opined that, because most obvious payoffs is office *i.e.*, cabinet portfolios offers administrative control over policy output and votes in policy decisions at cabinet meetings. They further argue that, policy positions are important elements in the party competition to achieve office, and that even an office seeking party receives payoff from the policies with which it is associated.

From the above theories of coalition, it is clear that, coalitions are formed for the sake of some rewards, material or immaterial. The partners combine together to win the game in order to have material reward for their labours. However, it is not possible for every partner to gain materially under all circumstances; gains may also be of a psychic nature. There may be a situation when a party is willing to "forgo material reward for the sake of obtaining the psychic reward of leadership".[13] Secondly, if the partners are more than two, some of them may go to the length of behaving like negotiators, whose concern is to draw advantage out of the obtaining situation. The underlying principles of a coalition system stand on the simple fact of temporary conjunction of specific interests. As a result of tugs and pulls, a point of equilibrium is arrived at where the actors agree to lay down their arms to have their united strength for the realization of the goal, however limited it may be.

15.5 Formation of Coalition Governments in Manipur

As mention above, the making and breaking of coalition governments in Manipur is probably one of the highest in India. On February 23, 1990 the 'United Legislature Front'[14] formed the government under the leadership of MPP stalwart R. K. Ranbir Singh.[15] This was the first non-Congress government in Manipur after more than a decade. There were several factors contributing to the formation of the *United Legislature Front* (ULF) in Manipur. However, the most important factor had been the anti-Congressism wave during Assembly election held in 1989 across several states in the country which had its consequences in Manipur. Therefore, the anti-Congressism wave had temporarily united the non-Congress parties in Manipur. The Congress party in Manipur had been the single largest party by bagging 26 seats although it did not formed the government since no other party had supported the party; a justification of anti-Congressism in Manipur. Therefore, the formation of the ULF was not base on the foundation of socio-economic programmes rather it was built on the issue of anti-Congress stance which is clearly a non-political or non-socio-economic issue. The coalition which consists of six political parties with varying ideologies built in the absence of common ideological visions came to an end with the imposition of President's rule on January 7, 1992, as a result of massive disqualification of 21 legislators due to internal bickering within both the ruling ULF and the opposition to the Congress.[16] With the revocation of the President's rule, on April 7, 1992 a Congress led 'Joint Legislature Party' another coalition government under the leadership of R.K Dorendro Singh of Congress party was installed.[17] Unlike the previous coalition, the CLP was supported by 55 legislators in the 60 members House. Interestingly, majority of the non-Congress members who were unwilling to extend support to the Congress party earlier have changed their positions after splits, a large scale defections and renaming of parties. Astonishingly, arch rivals R. K. Dorendro Singh (Congress Party) and R.K Ranbir Singh (MPP) agreed to work together.[18] It may be note worthy here that Mr. Ranbir Singh was not given the opportunity to prove majority support or lost of confidence in the House by the then governor of Manipur V.K. Nayar; another instance of Constitutional farce. Another responsible factor was the installation of the Congress government at the Centre in 1991 supported by AIADMK and other smaller parties.[19]

Thus, the impact on the state politics can be seen from the unruly behavior of the legislatures of Manipur. As many as 21 legislators were disqualified under the provision of Anti-defection law of the Indian Constitution for floor crossing and horse-trading since Congress party re-captures power at the Centre after Lok Sabha election in 1991.[20] The ministry lasted for a year before President's rule was imposed on December 31, 1993 due to bad law and order condition in the state.[21] On December 14, 1994 another Congress led coalition ministry was installed with Rishang Keishing as the Chief Minister supported by Janata Dal-10, Congress (s)-6, and Congress (I)-26 legislatures. The re-instatement of Rishang Keishing as the Chief Minister led to the resignation of the then Governor of Manipur V. K. Nayar on December 22, 1994. Earlier, the Governor V.K. Nayar in his report had held Rishang Keishing guilty of inciting the Tangkhuls against the Kukis in the ongoing ethnic feud.[22] The Governor was also not consulted for the re-instatement of Rishang Keishing as the Chief Minister which was itself a blow to the spirit of the Constitution of India.[23] This was another instance when the government at the Centre bulldozed its way into power in Manipur.

On February 25, 1995 Rishang Keishing led JLP ministry was again installed. The Governor O. N. Shrivastava's helplessness to accept the intimidation of the Congress high command to invite the party led by Rishang Keishing led to the defection of 16 legislators from opposition front who had change side by splitting, defection and renaming of parties.[24] Therefore, it was the habitual defectors and intimidation by the Congress high command which had enabled the Congress party to gradually increase its strength from 22 MLAs to 38 members of the JLP. Consequently, the coalition built by means of maneuvering defection shatters in the same manner when the non-Congress parties recapture power at the Centre during the month of December in 1997when 23 legislators of CLP split the party leading to the formation of 'Manipur State Congress Party'. The next coalition ministry under the banner of 'United Front' led by MSCP was installed on December 16, 1997. The coalition comprises of Manipur State Congress Party-23 (splinter group of Congress), MPP-11, FPM-2, CPI-1 and one Independent.[25] The formation of United Front ministry can be taken as a result of change of guard at the Centre though there were regional factors of immediate concerns. The ministry consist of 34 ministers which was the biggest ever constituted in the history of Manipur Assembly. The United Front parties recaptured power at the Centre after the Lok Sabha election held in 1996 under the leadership of Atal Behari Vajpayee on 16th May 1996, just before it was taken over by H.D. Deve Gowda and later by I.K. Gujral (May 1, 1996 –April 21, 1999).[26] The installation of the non-Congress government was immediately followed by political manipulation in Manipur. For example, Radhabinod Koijam (Deputy Chief Minister), N. Biren Singh (Sericulture Minister), Nimaichand Luwang (Higher Education Minister) and S. Rajen Singh MLA resigned immediately with the formation of *United Front ministry* at the Centre. The resignation of three prominent Congressmen was an indication of consequences of change of guard at the Centre. The UF ministry completed the remaining tenure of the Sixth Assembly. The seventh Assembly election in Manipur witnessed the formation of coalition of regional political parties. The period was marked with sensitive regional issues such as protection of integrity of Manipur in the wake of Indo-Naga cease fire.[27] Nipamacha Singh led United Democratic Alliance comprises of MSCP and Federal Party of Manipur with 29 members. Later, the UDA was renamed as United Front of Manipur with the defection of ten members from the opposition front. The UFM ministry under the leadership of W. Nipamacha Singh was installed on February 25, 2000. Like in the past, the Nipamacha Singh's Ministry was a short-lived one due to the political tussle between the Speaker Dhananjoy and the Chief Minister. When Nipamacha Singh lost confidence of the House, Radhabinod Koijam leader of the short-lived *United Democratic Alliance* formed the next coalition ministry during the month of January 2001, which like the previous coalition ministries collapsed on May 21, 2001. The collapse of the ministry only brings another imposition of President's rule in the state.

15.6 Conclusion

The study of the coalition formation in 1990 reveals that it was the anti-Congress wave that brought the *United Legislature Front* (ULF) coalition partners together rather than having ideological affinities as a cohesive force. Similarly, the coalition government formed on April 7, 1992 by the two archrival parties namely congress (I) and MPP was purely an opportunistic alliance to gain power. The other non-Congress (I) parties namely CPI, Janata Dal and Kuki National Assembly, which were not willing to form government with the Congress (I) earlier, also supported the ministry. Another Congress (I) led coalition ministry led by Rishang Keishing installed on December 14, 1994 enjoyed the support of the Janata Dal with 10 members and Congress (s) with six members. The re-instatement of Rishang as the Chief Minister without consulting the Governor, intimidation by the Prime Minister in affairs of the state in formation of the coalition governments in 1994 and 1995 shows that the government at the Centre had bulldoze its way into power in Manipur. The allocation of ministerial berths for 30 to 34 members to keep the coalition government intact by Nipamacha Singh in 1997 and in 2000 had proved that 'office seeking is more important than policy seeking' in the formation of coalition governments in Manipur. Policy pursuit models which presumed that coalition governments are easily formed when parties are not at a policy distance from each other had not been an important factor for the formation of the coalition governments in Manipur. But parties and individual legislators in Manipur are more interested in the pursuit of power or office in total disregard of ideological considerations. 'Power or office is what matters most' to the members of Manipur politicians. Thus, the essence of the politics of coalition formation in Manipur is opportunism and self-interest, whether it's a group or individuals. This has been the guiding tenets of the coalition governments in Manipur.

Thus, the formation of the coalition governments in Manipur during the study period suggest that none of the coalition formations were policy based but rather a bunch of parties with varying ideologies and party affiliations. Securing power and being into office is seen as one determination factor for the formation of the coalition governments. Frequent shift of party loyalty, so as not to miss out any opportunity to be part of office occupants took place almost on a regular basis. Coalitions are formed with different names by the same people responsible for breaking the coalition. This type of coalition formation had been possible because of leadership crisis, dissatisfaction over the allocation of portfolios or distribution of ministerial berths among the coalition partners. Therefore, one can conclude that coalitions are formed for the sake of sharing power amongst diversified interest groups and individual without a common ideological platform, based on long or short term policies and programmes. Therefore, the frequency of forming coalition governments leads us to assume that the coalition governments were formed for the sake of conveniences or a marriage for convenience. The frequency of coalition formation endures the fact that political parties were merely office seekers or to secured power without putting much weight on long or short term policies and programmes for the development of the state. Thus, the deep root cause of instability or unstable governments had been seen in the nature and patterns of office or power seekers resulting in the frequent formation of the coalition governments in Manipur.

Endnotes

1. R.P. Singh., *Electoral Politics in Manipur: A Spatio-Temporal Study*. New Delhi: Concept Publishing Company, 1981, pp. 22-23.

2. Gregory M. Luebbert., "Coalition Theory and Government Formation in Multi-party Democracies", in *Comparative Politics*, Jan. 1983, p. 246.

3. William H. Riker. *The Study of Coalitions*, New Haven: Yale University, 1962. p.524.

4. Lane and Ersson. *European Politics: An Introduction*. New Delhi: Sage Publications, 1996, p. 137.

5. Michael Laver & Ian Budge. "Office Seeking and Policy Pursuit in Coalition Theory", *Legislative Studies Quarterly*, XI, No. 4, November 1986, p. 494.

6. Abram De Swaan., *Coalition Theories and Cabinet Formation*. Amsterdam: Elsevier, 1973. p.88.

7. Michael Leiserson, "Factions and Coalitions in one-party Japan: An Interpretation Based on Theory of Games", *American Political Science Review*, September 1968, pp. 770-87.

8. Robert Axelrod, *Conflict of Interests: A theory of divergent goals with application of politics*. Chicago: Markham, (1970); Abraham De Swaan., *Op.cit.,*(1973).

9. Wolfgang C, Muller & Kaare Strom., (ed.) (1999). *Policy, Office or Votes: How Political Parties in Western Europe make Hard Decision.* Cambridge: Cambridge University Press, p. 5.

10. Norman Schofield & Michael Laver. 'Bargaining Theory and Portfolio Payoffs in European Coalition Governments 1945-83', *British Journal of Political Science,* Vol. 15, No.2 (Apr.1985) p.40.

11. *Ibid.*

12. *Ibid.*

13. M. Laver, & Ian Budge., (eds.), *Op.cit.,* (1992), p. 486.

14. W.H. Riker., *Op.cit.,* (1962), p. 24.

14. Manipur Peoples' Party, Janata Dal, Congress (S), Kuki National Assembly, National Peoples' Party, Communist Party of India.

15. *Manipur Mail* (Imphal), February 24, 1990, p.1.

16. MPP leading partner of the ULF splits, one faction led by the Chief Minister R.K Ranbir Singh while the party President Bhuban Singh led the other faction. Congress (I) was split by I. Tompok Singh along with 14 Congressmen and formed a new party called, 'Manipur Congress'.

17. *Ibid.*

18. "Manipur Developments: Making a mockery of Anti-defection Act", in *Indian Expres,* New Delhi, May 3, 1992, p. 1

19. S. Ranjan Raj., Coalition politics in India: Dimensions of federal power sharing, New Delhi: Manak Publications Pvt. Ltd., 2009, p. 118.

20. Iboyaima Laithangbam., "Political chameleons strengthen Keishing", in *Eastern Panorama,* November 1995, p. 17.

21. Kuki-Naga ethnic conflict in the Hill areas of Manipur claims 321 lives in another communal clash between the Meiteis and Pangals (Manipuri Muslims) in Imphal areas 94 lives were lost besides injuring more than 500 persons.

22. "Old games in Manipur", in *Statesman* (Calcutta), September 26, 1994, p. 6.

23. "Manipur Happenings", in *Tribune* (Chandigarh), January 24, 1995, p.1.

24. The 16 defectors merge with Congress party during the month of July 1995 after meeting with the Prime Minister P.V. Narasimha Rao.

25. Indian Recorder, (Vol. V, No. 4), 1998, p. 3389.

26. S. Ranjan Raj., *Op.cit.,* p. 110.

27. Indo-Naga ceasefire was signed in 1997 without territorial limit includes three districts (Chandel, Ukhrul and Tamenglong) of Manipur in the propose Greater Nagaland policy of NSCN (IM).

References

Abram De Swaan. 1973. *Coalition Theories and Cabinet Formation.* Amsterdam: Elsevier.

Gregory M. Luebbert. 1983. 'Coalition Theory and Government Formation in Multi-party Democracies', in *Comparative Politics*. Vol. 15 No. 2, Jan., p. 235-249.

Iboyaima, Laithangbam. 1995. 'Political chameleons strengthen Keishing', in *Eastern Panorama*. November, (p.17).

Lane and Ersson. 1996. *European Politics: An Introduction.* New Delhi: Sage publications.

Michael Laver & Ian Budge.1986. 'Office Seeking and Policy Pursuit in Coalition Theory', in *Legislative Studies Quarterly.* Vol. XI, No. 4, November (p. 494).

Michael Leiserson. 1968. 'Factions and Coalitions in one-party Japan: An interpretation based on theory of games', in *American Political Science Review.* September (pp. 770-87).

Norman Schofield & Michael Laver. 1985. 'Bargaining Theory and Portfolio Payoffs in European Coalition Governments 1945-83', in *British Journal of Political Science*, Vol. 15(2) April (p. 40).

Raj, S. Ranjan. 2009. *Coalition politics in India: Dimensions of federal power sharing.* New Delhi: Manak Publications Pvt. Ltd.

Robert Axelrod. 1970. *Conflict of Interests: A theory of divergent goals with application of politics,* Chicago: Markham, *Op.cit.,*(1973) Abraham De Swaan.

Singh, R.P. 1981. *Electoral Politics in Manipur: A Spatio-Temporal Study.* New Delhi: Concept Publishing Company.

William, H. Riker. 1962. *The study of coalitions.* New Haven: Yale University.

Wolfgang C, Muller & Kaare Strom., (eds.). 1999. *Policy, Office or Votes: How Political Parties in Western Europe make Hard Decision.* Cambridge: Cambridge University Press.

Manipur Mail. 1990. Imphal, February 24, (p.1).

Indian Express. 1992. 'Manipur Developments: Making a mockery of Anti-defection Act', New Delhi. May 3, (p.1).

Statesman (Calcutta). 1994. 'Old games in Manipur', September 26 (p.6).

Tribune (Chandigarh). 1995. 'Manipur Happenings'. January 24 (p.1).

Indian Recorder. 1998. Vol. V, No. 4 (p. 3389).

About the Author

P. Khongsai is currently pursuing his Post-Doctorial from Manipur University. He has published a research paper in an edited book, The Kuki Society: Past, Present, and Future (2011) titled as: Dynamics of coalition politics for social change. He has worked as a Research Investigator in the Department of Philosophy under UPE- Areas Studies in North-eastern Hill University. He is a lifetime member of Indian Journal Political Science Association and is Coordinator of Kuki Research Forum.

E-mail: psailenthang@gmail.com   Mobile: 9612125529

Index